The PROACT® Root Cause Analysis
Quick Reference Guide

The PROACT® Root Cause Analysis
Quick Reference Guide

Kenneth C. Latino, Mark A. Latino, and Robert J. Latino

CRC Press
Taylor & Francis Group
Boca Raton   London   New York

CRC Press is an imprint of the
Taylor & Francis Group, an **informa** business

First edition published 2020
by CRC Press
6000 Broken Sound Parkway NW, Suite 300, Boca Raton, FL 33487-2742

and by CRC Press
2 Park Square, Milton Park, Abingdon, Oxon, OX14 4RN

© 2021 Taylor & Francis Group, LLC

CRC Press is an imprint of Taylor & Francis Group, LLC

Library of Congress Cataloging-in-Publication Data
Names: Latino, Kenneth C., author. | Latino, Mark A., author. |
Latino, Robert J., author.
Title: The PROACT® root cause analysis : quick reference guide / Kenneth C.
Latino, Mark A. Latino, and Robert J. Latino.
Description: First edition. | Boca Raton, FL : CRC Press, 2020. | Includes
bibliographical references and index.
Identifiers: LCCN 2020015736 (print) | LCCN 2020015737 (ebook) |
ISBN 9780367517380 (hardback) | ISBN 9781003055013 (ebook)
Subjects: LCSH: Root cause analysis.
Classification: LCC TA169.55.R66 L38 2020 (print) | LCC TA169.55.R66
(ebook) | DDC 658.4/013—dc23
LC record available at https://lccn.loc.gov/2020015736
LC ebook record available at https://lccn.loc.gov/2020015737

ISBN: 978-0-367-51738-0 (hbk)
ISBN: 978-1-003-05501-3 (ebk)

Typeset in Times
by codeMantra

CONTENTS

PREFACE

What is Root Cause Analysis (RCA)? It seems like such an easy question to answer, yet from novices to veterans and practitioners to providers, we cannot seem to agree (nor come to consensus) on an acceptable definition for the industry. Why? We will discuss our beliefs as to why in this mini book.

For full disclosure, this mini book is just that, a book for those who want to skip the fluff and get right to the meat. In this book, we do not get into much theory, we go straight into practice. This is a "how to" book plain and simple. There are also some publishing constraints you should be aware of when one writes a mini book; **we are limited to only ten figures/tables for the whole book!** So, we had to be pretty selective about which graphics had the greatest impact for the reader and therefore would be included in the book proper. However, fear not, there's always a workaround. For figures and tables that did not make the cut, they can be found online at www.routledge.com/9780367517380. We will put things such as forms, templates, and guidance documents in this location.

We tried to write this text in a conversational style because we believe this is a format that most "rooticians" can relate to. Basically, we wrote like we were teaching a workshop.

Readers will find that much of our experience comes not only from the practicing of RCA in the field but more so from our experiences with the over 10,000 analysts whom we have taught and mentored over the years. Additionally, we participate in many online discussion forums where we interact with beginners, veterans, and most providers for the betterment of the RCA field.

So as you can see, we try to bring many diverse perspectives to the table, while making the pursuit of RCA a practical one, not a complex one. We certainly want to avoid falling into the "paralysis-by-analysis" trap when looking at something like RCA—that would be hypocritical—would it not?

There are many RCA methodologies on the market, and we discuss them in generalities so as not to put the microscope on any individual or proprietary approach. In this manner, we can discuss the pros and cons of each type of approach, and readers can decide the level of breadth and depth that they require in their analysis.

Will everybody who reads this text agree with its content? No. Can they benefit regardless? Yes. We hope to spark debate within the minds of our readers where the differences are contrasted between how we approach RCA and how they are currently conducting them at their facilities.

Perhaps we will sway some to agree with certain premises in this text, and others will improve upon their current approaches with the ideas presented. Either way, the journey of the learning is what is most important. Analysts will collect the necessary data, sift out the facts, and make their own determination as to what they believe is best for them.

Robert J. Latino
Kenneth C. Latino
Mark A. Latino

AUTHORS

Kenneth C. Latino is President of Prelical Practical Reliability Solutions (www.prelical.com) in Daleville, VA. He has a bachelor of science degree in computerized information systems from Virginia Commonwealth University. He began his career developing and maintaining maintenance software applications in the continuous process industries. After working with clients to help them become more proactive in their maintenance activities, he began consulting and teaching industrial plants how to implement reliability methodologies and techniques to help improve the overall performance of plant assets.

Over the past few years, a majority of Kenneth's focus has centered around developing reliability approaches with a heavy emphasis on Root Cause Analysis (RCA). He has trained thousands of engineers and technical representatives on how to implement a successful RCA strategy at their respective facilities. He has co-authored two RCA training seminars: one for engineers and another for hourly personnel.

Kenneth is also co-software designer of the RCA program entitled the PROACT® Suite. PROACT® was a National Gold Medal Award winner in *Plant Engineering's* 1998 and 2000 Product of the Year competition for its first two versions on the market. He is currently president of the Practical Reliability Group, a Reliability consulting firm dedicated to delivering approaches and solutions that can be practically applied in any asset-intensive industry.

Mark A. Latino Mark came to RCI after 19 years in corporate America. During those years, he acquired a wealth of reliability, maintenance, and manufacturing experience. He worked for

Weyerhaeuser Corporation in a production role during the early stages of his career. He had an active part in Allied Chemical Corporation's (now Honeywell) Reliability Strive for Excellence initiative that was started in the 1970s to define, understand, document, and live the Reliability culture until he left in 1986. Mark spent 10 years with Philip Morris primarily in a production capacity that later ended in a Reliability Engineering role. Mark is a graduate of Old Dominion University and has a bachelor's degree in business that focused on production and operations management.

Robert J. Latino is CEO of Reliability Center, Inc. (RCI). RCI is a reliability consulting firm specializing in improving equipment, process, and human reliability. He received his bachelor's degree in business administration and management from Virginia Commonwealth University.

Robert has been facilitating RCA and Failure Mode and Effects Analysis (FMEA) with his clientele around the world for over 35 years and has taught over 10,000 students in the PROACT® methodology. He is co-author of numerous seminars and workshops on FMEA and RCA as well as co-designer of the award-winning PROACT® Suite Software Package.

Robert is a contributing author of *Error Reduction in Healthcare: A Systems Approach to Improving Patient Safety* and *The Handbook of Patient Safety Compliance: A Practical Guide for Health Care Organizations*.

Robert has also published a paper entitled, "Optimizing FMEA and RCA Efforts in Healthcare" in the *ASHRM Journal* and presented a paper entitled, "Root Cause Analysis Versus Shallow Cause Analysis: What's the Difference?" at the ASHRM 2005 National Conference in San Antonio, Texas. He has been published in numerous trade magazines on the topic of reliability, FMEA, and RCA and is also a frequent speaker on the topic at domestic and international trade conferences.

Robert has also applied the PROACT® methodology to the field of Terrorism and Counter Terrorism via a published paper entitled, "The Application of PROACT® RCA to Terrorism/Counter Terrorism Related Events."

INTRODUCTION TO THE FIELD OF ROOT CAUSE ANALYSIS AND PROACT®

WHAT IS ROOT CAUSE ANALYSIS (RCA)?

What a seemingly easy question to answer, yet no standard, generally accepted definition of Root Cause Analysis (RCA) exists in the industry today, of which we are aware. Technical societies, regulatory bodies, and corporations have their own definitions, but it is rare to find two definitions that match. For the sake of having an anchor or benchmark definition, we will use the definitions provided by the Department of Energy (DOE) guideline entitled "Root Cause Analysis Guidance Document (DOE-NE-STD-1004–92)."[1]

In the DOE document referenced above, the following is cited:

> The basic reason for investigating and reporting the causes of occurrences is to enable the identification of corrective actions adequate to prevent recurrence and thereby protect the health and safety of the public, the workers and the environment.

The document goes to say that every root cause investigation and reporting process should include the following five phases:

I. Data Collection
II. Assessment
III. Corrective Actions
IV. Inform
V. Follow-Up

When we look at any investigative occupation, these five steps are critical to the success of the investigation. As we progress through

---

[1] http://www.hss.energy.gov/nuclearsafety/ns/techstds/standard/nst1004/nst1004.pdf.

this text, we will align the steps of the PROACT® methodology with each of these steps in the DOE RCA process.

For the purposes of this text, while aligning with the DOE guideline, we will use our own definition of RCA, which is

> The establishing of logically complete, evidence-based, tightly coupled chains of factors from the least acceptable consequences to the deepest significant underlying causes.

This is a variation of a definition that was proposed on the RCA discussion forum at www.rootcauselive.com.[2]

While a seemingly complex definition, let's break down the sentence into its logical components and briefly explain each:

1. Logically Complete—This means all the options (hypotheses) are considered and either proven or disproven using hard evidence.
2. Evidence Based—This means hard evidence is used to support the hypotheses as opposed to using hearsay and treating it as fact.
3. Tightly Coupled Chains of Factors—These fancy words mean we are using cause-and-effect RCA approaches as opposed to categorical (cause categories where we brainstorm disconnected possibilities in each category) RCA approaches.
4. Least Acceptable Consequences—This is the point where the event that has occurred is no longer acceptable (a trigger of some sort has been hit) and an investigation is launched.
5. Deepest Significant Underlying Causes—These seemingly intimidating words mean at what point do we stop drilling down and determine going deeper adds no value to the organization?

This definition certainly encompasses and embodies the intent of the DOE guideline for RCA.

SUMMARY OF THE PROACT® INVESTIGATION MANAGEMENT SYSTEM

For the remainder of this text, we will follow the simple steps of the PROACT® Investigative Management System, consistent with the DOE guidelines. The PROACT® acronym stands for the following:

---

[2] This discussion forum is associated with www.rootcauselive.com and moderated by Mr. C. Robert Nelms.

1. **PR**eserving Evidence/Data
2. **O**rganizing the RCA Team
3. **A**nalyzing the Event
4. **C**ommunicate Findings & Recommendations
5. **T**racking Bottom-Line Results

As you read this text, think about how each step of the process applies to where you work. When we think about any investigative occupation, they all follow these steps to some degree. This should be our measure to determine if we are doing true RCA or simply Shallow Cause Analysis, where we succumb to taking process shortcuts due to time pressures imposed on us.

WHY DO UNDESIRABLE OUTCOMES OCCUR? THE BIG PICTURE

We must put aside the industry that we work in and follow along from the standpoint of the human being. In order to understand why undesirable outcomes exist, we must understand the mechanics of failure.

Virtually all undesirable outcomes are the result of human errors of omission or commission (or decision errors or choices, as they will be used interchangeably from now on). Experience in industry indicates that any undesirable outcome will have, on average, a series of 10–14 cause-and-effect relationships that queue up in a particular pattern for that event to occur.

This dispels the commonly held myth that one error causes the ultimate undesirable outcome. All such undesirable outcomes will have their roots embedded in the physical, human, and latent areas.

Physical Roots: are typically found soon after errors of commission or omission. They are the first physical consequences resulting from choices made. Physical roots, as will be described in detail in coming chapters, are tangible things we can see.

Human Roots: are decisions that are made that did not go as planned. These are the actions (or inactions) that trigger the physical roots to surface.

Latent Roots: are the organizational systems that are flawed in some manner (i.e., inadequate, insufficient, and/or nonexistent). These are the organizational support systems (i.e., procedures, training, incentive systems, purchasing habits) that are typically put in place to help our workforce make better decisions. Latent roots are the expressed intent of the human decision-making process.

WHAT ARE THE ELEMENTS OF A TRUE ROOT CAUSE ANALYSIS SYSTEM?

In order to recognize what is RCA and what is NOT RCA (Shallow Cause Analysis), we would have to define the criteria that must be met in order for a process and its tools to be called Root Cause Analysis. In the absence of a universally accepted standard, let's consider the following essential elements[3] of a true RCA process:

1. **Identification of the *Real* Problem to be Analyzed in the First Place:** About 80% of the time we are asked to assist on an investigation team, the problem presented to us is not the problem at hand.

2. **Identification of the Cause-and-Effect Relationships that Combined to Cause the Undesirable Outcome:** Being able to correlate deficient systems directly to undesirable outcomes is critical. Using categorical approaches (as we will explain in Chapter 4, "Analyzing the Data") will often yield fewer comprehensive results than cause-and-effect approaches.

3. **Disciplined Data Collection and Preservation of Evidence to Support Cause-and-Effect Relationships:** It is safe to say that if we are not collecting data to validate our hypotheses, we are not properly conducting a comprehensive RCA.

4. **Identification of All Physical, Human, and Latent Root Causes Associated with Undesirable Outcome:** If we are not identifying system deficiencies that lead to poor decision-making, then, again, we are not properly conducting a comprehensive RCA.

5. **Development of Corrective Actions/Countermeasures to Prevent Same and Similar Problems in the Future:** If we have merely developed good recommendations but never implement them, then we will not be successful in our RCA efforts. This is where the ball is often dropped as well-intentioned people are pulled away by reactive work, and these proactive opportunities fall by the wayside.

6. **Effective Communication to Others in the Organization of Lessons Learned from Analysis Conclusions:** One of the greatest benefits of a successful RCA is the dissemination of

---

[3] Latino, Robert J. PROACT® Approach to Healthcare Workshop. January 2005. www.proactforhealthcare.com.

the lessons learned to avoid recurrence elsewhere in the organization. Oftentimes, successful analyses end up in a paper filing system only to be suppressed from those who could benefit from the lessons learned in the analysis.

This mini book is intended to streamline the understanding of the PROACT® RCA process for those that have to do the work. So, we will avoid as much theory as possible and outline the basics associated with "how to" conduct more, better, and faster (more efficient) RCAs. Let's get to it!!

2

PRESERVING EVENT DATA

THE PROACT® RCA METHODOLOGY

The term "proact" has recently come to mean the opposite of react. This may seem to conflict with PROACT®'s use as a Root Cause Analysis (RCA) tool. Normally, when we think of RCA, the phrase "after-the-fact" comes to mind—after, by its nature, an undesirable outcome that must occur in order to spark action. So how can RCA be coined as proactive?

RCA tools can be used in a reactive fashion and/or a proactive fashion. The RCA analyst will ultimately determine this. When we use RCA only to investigate *incidents* that are defined by regulatory agencies, we are responding to the daily needs of the field. This is strictly reactive.

However, if we apply these RCA principles proactively, we will uncover events that many times are not even recorded in our Computerized Maintenance Management Systems (CMMSs) or the like. This is because such events happen so often that they are no longer anomalies. They are a part of the job. They have been absorbed into the daily routine. By uncovering such events and analyzing them, we are being proactive because unless we look at them, no one else will.

THE GREATEST BENEFITS FROM PERFORMING RCAS WILL COME FROM THE ANALYSIS OF CHRONIC EVENTS, thus using RCA in a proactive manner. We must understand that oftentimes we get sucked into the "paralysis by analysis" trap and end up expending too many resources to attack an issue that is relatively unimportant when considering the big picture. We also at times refer to these as the "political-failures-of-the-day." Trying to do RCA on everything will destroy a company. It is overkill, and companies do not have the time or resources to do it effectively.

RCA TEAM PROCESS FLOW DIAGRAM

As we tell our students, we provide the architecture of an RCA methodology. It will not work the same in every organization. The model

6

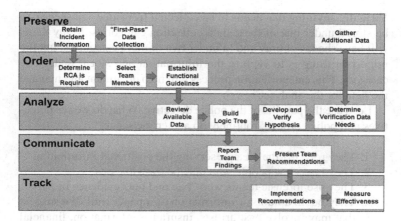

Figure 2.1 PROACT® RCA process flowchart.

or framework should be molded to each culture into which it is being forced. We must all play with the hand we are dealt. We do the best we can with what we have.

To that end, the process flow involved with such team activities might look like that presented in Figure 2.1.

We can speak ideally about how RCA teams should function, but rarely are there ideal situations in the real world.

THE 5-P's DATA COLLECTION CONCEPT

*Pr*eserving Failure Data is the *PR* in PROACT®. In a typical high-profile RCA, an immense amount of data is typically collected and then must be organized and managed.

Consider this scenario: a major upset just occurred in your facility. You are charged to collect the necessary data for an investigation. What is the necessary information to collect for an investigation or analysis? We use a 5 P's approach, where the P's stand for the following:

1. **P**arts
2. **P**osition
3. **P**eople
4. **P**aper
5. **P**aradigms

Virtually anything that needs to be collected from an event scene can be stored under one of these data collection categories. Many items will have shades of gray and fit under two or more categories, but the important thing is to capture the information and slot it under one category. This categorization process will help document and manage the data for the analysis.

Let's use the parallel of the police detective. What do we see detectives and police officers routinely do at a crime scene?

1. They rope off the area preserving the positional information.
2. They interview people who may be eyewitnesses.
3. They use forensic teams who "bag and tag" evidence or parts.
4. They begin a hunt for information or a paper trail of a suspect that may involve past arrests, insurance information, financial situation, etc.
5. They develop theories (hypotheses) about the situation such as "… he was always at home during the day and away at night. We would see children constantly visiting for five minutes at a time. We think he is a drug dealer." These are the paradigms that people have about situations that are important, because if they truly believe these paradigms, then they are basing their decisions on them. This can be dangerous.

PARTS

Parts will generally mean something physical or tangible. The potential list is endless, depending on the facility where the RCA is conducted. For a rough sampling of what is meant by parts, review the following lists for applicability to your work environment:

Continuous process industries (oil, steel, aluminum, paper, chemicals, etc.)
Bearings
Seals
Couplings
Impellers
Bolts
Flanges
Grease samples
Product samples
Water samples

Tools
Testing equipment
Instrumentation
Tanks
Compressors
Motors

Discrete product industries (automobiles, package delivery, bottling lines, etc.)
Product samples
Conveyor rollers
Pumps
Motors
Instrumentation
Processing equipment

Healthcare (hospitals, nursing homes, outpatient care centers, long-term care facilities, etc.)
Medical diagnostic equipment
Surgical tools
Gauze
Fluid samples
Blood samples
Biopsies
Medications
Syringes/needles
Testing equipment
IV pumps
Patient beds/rails

This is just a sampling to give you a feel for the type of information that may be considered under the parts category.

POSITION

Positional Data Related to Physical Space

Positional data is the least understood and what we consider to be the most important. Positional data comes in the form of two different dimensions, one being physical space and the other being point in time. Positions in terms of space are vitally important to an analysis because of the facts that can be deduced.

Bear with us on this reflection using two memorable failures; the space shuttles Challenger and Columbia. We use these to demonstrate the value of positional data to any investigation.

When the space shuttle Challenger exploded on January 28, 1986, it was approximately 5 miles in the air. Films from the ground provided millisecond-by millisecond footage of the parts that were being dispersed from the initial cloud. From this positional information, trajectory information was calculated, and search and recovery groups were assigned to approximate the locations of where vital parts were located. Approximately 93,000 square miles of ocean were involved in the search and recovery of shuttle evidence in the government investigation.[1] While this is an extreme case, it shows how position information is used to determine, among other things, force.

While on the subject of the shuttle Challenger, other positional information that should be considered is:

1. "Why was it the right Solid Rocket Booster (SRB) and not the left?"
2. "Why was it the aft (lower) field joint attachment that leaked versus the upper field joint attachment?"
3. "Why was the leak at the O-ring on the inside diameter of the SRB versus the outside diameter?"
4. These are questions regarding positional information that had to be answered.

Positional Data Related to Time

Now let's look at positions in time and their relative importance. Monitoring *positions in time* in which undesirable outcomes occur can provide information for correlation analysis. By recording historical occurrences, we can plot trends that identify the presence of certain variables when these occurrences happen.

Let's look at the shuttle Challenger again. Most of us remember the incident and the conclusion reported to the public: an O-ring failure resulting in a leak of solid rocket fuel.

However, if we look at the positional information from the standpoint of time, we would learn that the O-rings had evidence of

---

[1] Lewis, Richard S. 1988. *Challenger: The Final Voyage*. New York: Columbia University Press.

secondary O-ring erosion on 15 of the previous 25 shuttle launches.[2] When the SRBs are released, they are parachuted into the ocean, retrieved, and analyzed for damage. The correlation of these past launches that incurred secondary O-ring erosion showed that low temperatures were a common variable. The *positions in time* information aided in this correlation.

Ironically, in the shuttle Columbia breakup on January 16, 2003, there were seven occurrences of bipod ramp foam events (missing tiles) since the first mission STS-1.

The long and short of it is that the loss of foam tiles from the main fuel tanks and their subsequent impact on the shuttle vehicle were not a new phenomenon—just like the O-ring erosion occurrences. Collecting the positions in time of these occurrences and mapping them out on a timeline prove these correlations.

Moving into more familiar environments, we can review some general or common positional information to be collected at almost any organization:

1. Physical position of parts at scene of incident
2. Point in time of current and past occurrences
3. Position of instrument readings
4. Position of personnel at time of occurrence(s)
5. Position of occurrence in relation to overall facility
6. Environmental information related to position of occurrence such as temperature, humidity, and wind velocity.

We are not looking to recruit artists for these maps or sketches. We are simply seeking to ensure that everyone sees the situation the same way based on the facts at hand. Again, this is just a sampling to get individuals in the right frame of mind of what we mean by positional information.

PEOPLE

The "People" category is the more easily defined "P." This is simply who we need to initially talk to in order to obtain information about an event. The people we must talk to first should typically be the physical observers or witnesses to the event. Efforts to obtain such interviews should be relentless and immediate. We risk the chance

---

[2] Lewis, Richard S. 1988. *Challenger: The Final Voyage*. New York: Columbia University Press.

of losing direct observation when we interview observers' days after an event occurs. We will ultimately lose some degree of short-term memory and risk the observers having talked to others about their opinion of what happened. Once observers discuss such an event with another outsider, they will tend to reshape their direct observation with the new perspectives.

We have always identified the goal of an interview with an observer to be that we must be able to see through their eyes what they saw at the scene. The description must be vivid, and it is up to the interviewer to obtain such clarity through the questioning process.

Interviewing skills are necessary in such analytical work. People must feel comfortable around an interviewer and not intimidated. A poor interviewing style can ruin an interview and subsequently an analysis or investigation. A good interviewer will understand the importance and value of body language. Experts estimate approximately 55%–60% of all communication between people is through body language. Approximately 30% of communication is through the tonal voice and 10%–15% is through the spoken word.[3] This is very important when interviewing because it emphasizes the need to interview in person rather than over the telephone. If you look at the legal profession, lawyers are professionals at reading the body language of their clients, their opposition, and the witnesses. Body language clues will direct their next move. This should be the same for interviews associated with an undesirable outcome. The body language will tell interviewers when they are getting close to information they desire, and this will direct the line and tone of subsequent questioning.

Consider another profession that we might not think of as having a strong relationship with body language—professional poker players. It does not take the novice long to realize the strength of the cards you are dealt does not determine if you are a winner. Professional poker players play their hands based on their read of the body language of their opponents. They know there are certain involuntary responses of the body by certain players that indicate they are holding a strong hand or they are likely bluffing. This further validates the importance and effect of body language when interviewing.

When interviewing during an RCA, it is also important to consider the logistics of the interview. Where is the appropriate place to interview? How many people should we interview at a time? What types of people should be in the room at the same time? How will we record

[3] Lyle, Jane. 1990. *Body Language*. London: The Hamlyn Publishing Group Limited.

all the information? Preparation and environment are very important factors to consider.

We have the most success in interviews when the interviewees are from various departments and more specifically from different "kingdoms" or silos. We define kingdoms as entities that build their castles within facilities and tend not to communicate with each other. Examples can be maintenance versus operations, labor versus management, doctors versus nurses, and hourly versus salary. When such groups get together, they learn a great deal about the others' perspective and tend to earn a respect for each other's position. This is another added benefit of an RCA—people start to meet and communicate with others from different levels and areas.

If an interviewer is fortunate enough to have an associate analyst to assist, the associate analyst can take the notes while the interviewer focuses on the interview. It is not recommended that recording devices be used in routine interviews as they are intimidating, and people believe the information may be used against them later. In some instances, where significant legal liabilities may be at play, legal counsel may impose such actions. However, if they do, they are generally doing the interviewing. In the case of most chronic failures or events, such extremes are rare.

Typical people to interview will again be based on the nature of the industry and the event being analyzed. As a sample of potential interviewees, consider the following list:

1. Observers
2. Maintenance personnel
3. Operations personnel
4. Management personnel
5. Administrative personnel
6. Technical personnel
7. Purchasing personnel
8. Storeroom personnel
9. Vendor representatives
10. Original equipment manufacturers (OEMs)
11. Personnel at other similar sites with similar processes
12. Inspection/quality control personnel
13. Risk/safety personnel
14. Environmental personnel
15. Lab personnel
16. Outside experts

As stated previously, this is just to give you a feel for the variety of people who may provide information about any given event.

PAPER

Paper data is probably the most *understood* form of data. Being in an information age where we have instant access to data through our communications systems, we tend to be able to amass a great deal of paper data. However, we must make sure that we are not collecting paper data for the sake of developing a big file. Some companies seem to feel they are getting paid based on the width of the file folder. We must make sure the data we are collecting is relevant to the analysis at hand.

Keep in mind our detective scenarios discussed earlier and the fact they are always preparing a solid case for court. Paper data is one of the most effective and expected categories of evidence in court. Solid, organized documentation is the key to a winning strategy.

Typical paper data examples are as follows:

1. Chemistry lab reports
2. Metallurgical lab reports
3. Specifications
4. Procedures
5. Policies
6. Financial reports
7. Training records
8. Purchasing requisitions/authorizations
9. Nondestructive testing results
10. Quality control reports
11. Employee file information
12. Maintenance histories
13. Production histories
14. Medical histories/patient records
15. Safety records information
16. Internal memos/e-mails
17. Sales contact information
18. Process and instrumentation drawings
19. Past RCA reports
20. Labeling of equipment/products
21. Distributive Control System (DCS) strips
22. Statistical Process Control/Statistical Quality Control Information (SPC/SQC)

PARADIGMS

Paradigms have been discussed throughout this text as a necessary foundation of understanding how our thought processes affect our problem-solving abilities. But exactly what are paradigms? We will base the definition we use in RCA on futurist Joel Barker's definition, as follows:

> A paradigm is a set of rules and regulations that: 1) Defines boundaries; and 2) tells you what to do to be successful within those boundaries. (Success is measured by the problems you solve using these rules and regulations.)[4]

This is basically how groups of individuals view the world and react and respond to situations arising around them. This inherently affects how we approach solving problems and will ultimately be responsible for our success or failure in the RCA effort.

Paradigms are a by-product of interviews carried out in this process. Paradigms are recognizable because repetitive themes are expressed in these interviews from various individuals. How an individual sees the world is a mindset. When a certain population shares the same mindset, it becomes a paradigm. Paradigms are important because even if they are false, they represent the beliefs in which we base our decision-making. Therefore, true paradigms represent reality to the people that possess them.

Following is a list of common paradigms we see in our travels. We are not making a judgment as to whether they are true but rather they affect judgment in decision-making.

1. We do not have time to perform RCA.
2. We say Safety is number one, but when it comes down to brass tacks on the floor, cost is really number one!
3. This is impossible to solve.
4. We have tried to solve this for 20 years.
5. It's old equipment; it's supposed to fail.
6. We know because we have been here for 25 years.
7. This is another program-of-the-month.
8. We do not need data to go through this RCA process because we know the answer.
9. This is another way for management to "witch-hunt."

---

[4] Barker, Joel. 1989. *Discovering the Future: The Business of Paradigms.* Elmo, MN: ILI Press.

10. Failure happens; the best we can do is sharpen our response.
11. RCA will eliminate maintenance jobs.
12. It is a career-limiting choice to contradict the doctor (a nurse's perspective).
13. We fully trust the hospitals to be responsible for our care.
14. Hospitals are safe havens for the sick.
15. What we get is what we order; there is no need to check.
16. RCA is RCA; it is all the same.
17. We don't need RCA; we know the answer.
18. If the failure is compensated for in the budget, it is not really a failure anymore.
19. RCA is someone else's job, not mine.

Many of these statements may sound familiar. But think about how each statement could affect problem-solving abilities. Consider the following if-then statements:

1. If we see RCA as another burden (and not a tool) on our plate, then we will not give it a high priority.
2. If we believe management values profit more than Safety, then we may rationalize at some time that bending the Safety rules is really what our management wants us to do.
3. If we believe that something is "impossible" to solve, then we will not solve it.
4. If we believe we have not been able to solve the problem in the past, then no one will be able to solve it.
5. If we believe that equipment will fail because it is old, then we will be better prepared to replace it.
6. If we believe RCA is the program-of-the-month, then we will wait it out until the fad goes away.
7. If we do not believe data collection is important, then we will rely on word-of-mouth and allow ignorance and assumption to penetrate an RCA as fact.
8. If we believe that RCA is a witch-hunting tool, then we will not participate.
9. If we believe failure is inevitable, then the best we can do is become a better responder.
10. If we believe that RCA will eventually eliminate our jobs, then we will not let it succeed.
11. If a nurse believes it is career limiting to contradict a doctor's order, then someone will likely die as a result of the silence.

12. If we believe the hospital is in total control of our care, then we will not question things that seem wrong.
13. If we believe hospitals are safe havens for the sick, then we are stating we are not responsible for our own safety.
14. If we believe what we get is what we order, then we will not ever inspect when we receive an order and just trust the vendor.
15. If we believe all RCAs are the same, then techniques like the 5 Whys will be considered as comprehensive and thorough as PROACT®.
16. If we believe we know all the answers, then RCA will not be valued.
17. If we believe unexpected failures are covered for in the budget, then we will not attempt to resolve those unexpected failures.
18. If we believe RCA is someone else's job, then we are indicating our safety is the responsibility of others and not ourselves.

The purpose of these "if-then" statements is to show the effect that paradigms have on human decision-making. When human errors in decision-making occur, it is the triggering mechanism for a series of other subsequent errors until the undesirable event surfaces and is recognized.

Now we must discuss how we get all this information. When an RCA team has been commissioned, a group of data collectors must be assembled to brainstorm what data will be necessary to collect in order to start the analysis.

This first team session is just that, a brainstorming session of data needs. This is not a session to analyze anything. The group must be focused on data needs and not be distracted by the premature search for solutions. The goal of this first session should not be to collect 100% of the data needed. Ideally, our data collection attempts should aspire to capture about 60%–70% of the necessary data. All the obvious surface data should be collected first and the most fragile data.

In terms of ranking the fragility of collecting these "P's," we see it as the following:

1. *People* and *Position* are tied for first (information to get the fastest). This is not an accident. As we discussed earlier in this chapter, the need to interview observers is immediate in order to obtain direct observation. *Positional* information is equally important because it is the most likely to be disturbed the

quickest. Therefore, attempts to get such data should be performed immediately.

2. *Parts* are second because if there is not a plan to obtain them, they will typically end up in the trash can.

3. *Paper* data is generally static apart from process or online production data (DCS, SPC/SQC). Such technologies allow for automatic averaging of data to the point that if the information is not retrieved within a certain time frame, it can be lost forever.

4. *Paradigms* are last because we wish we could change them faster, but modifying behavior and belief systems take more time.

One preparatory step for analysts should be to always have a data collection kit prepared. Many times, such events occur when we least expect it. We do not want to have to run around collecting a camera, plastic bags, etc. If it is all in one place, it is much easier to be prepared in a minute's notice. Usually good models are from other emergency response occupations such as doctor's bags, fire departments, police departments, and Emergency Medical Technicians (EMTs). They always have most of what they need accessible at any time. Such a bag (in general) may have the following items:

1. Caution tape
2. Masking tape
3. Plastic Ziploc® bags
4. Gloves
5. Safety glasses
6. Ear plugs
7. Adhesive labels
8. Marking pens
9. Digital camera w/spare batteries
10. Video camera (if possible)
11. Marking paint
12. Tweezers
13. Pad and pen
14. Measuring tape
15. Sample vials
16. Wire tags to ID equipment

This is, of course, a partial listing, and depending on the organization and nature of work, other items would be added or deleted from the list.

The form used to collect such data should include at a minimum, the following fields to fill, whether manually or in an automated RCA software system like the PROACT® Investigation Management System (www.reliability.com).

1. Data Type/Category—Which of the 5 P's this form is directed at is listed. Each "P" should have its own form.
2. Person Responsible—The person responsible for making sure the data is collected by the assigned date.
3. Data to Collect—During the 5 P's brainstorming session, list all data necessary to collect for each "P."
4. Data Collection Strategy—This space is for listing the plan of how to obtain the previously identified *data to collect*.
5. Date to Be Collected by—Date by which the data is to be collected and ready to be reported to team.

3

ORGANIZING THE ANALYSIS TEAM

When a sporadic/acute event typically occurs in an organization, an immediate effort is organized to form a task team to investigate "WHY" such an undesirable event occurred. What is the typical make-up of such a task team? We see the natural tendency of management to assign the "cream-of-the-crop" experts to both lead and participate on such a team. While well intended, there are some potential disadvantages to this thought process.

TALE OF A 'TEAM'

Let's paint a real-life scenario in a manufacturing setting (even though it could happen anywhere). A sulfur burner boiler fails due to tube ruptures. The event considerably impacts production capabilities when it occurs. Maintenance histories confirm that such an occurrence is chronic as it has happened at least once a year for the past 10 years. Therefore, Mean-Time-Between-Failure (MTBF) is approximately one per year. This event is a high priority on the mind of the plant manager, as it is impacting his facility's ability to meet corporate production goals and customer demand in a reliable manner. He is anxious for the problem to go away. He makes the logical deduction that if he has tubes rupturing in this boiler, then it must be a metal's issue. Based on this premise, he naturally would want to have his best people on the team. He assigns his top metallurgist as the team leader because he has been with the company the longest and has the most experience in the materials lab. On the team, he will provide the metallurgist the resources of his immediate staff to dedicate the time to solve the problem. Does this sound familiar? The logic appears sound. Why wouldn't this strategy work?

Let's review what typically happens next. We have a team of say five metallurgists. They are brainstorming all the reasons these tubes could be bursting. At the end of their session(s) they conclude that more exotic metals are required, and the tube materials should be changed in order to be able to endure the harsh atmosphere in which they operate. Problem solved! However, this is the same scenario that went on for the past 10 years and they kept replacing the tubes year after year and the tubes kept rupturing.

Think about what just went on with that team. Remember our earlier discussion about *paradigms* and how people view the world. How do we think the team of metallurgists view the world? They all share the same "box." They have similar educational backgrounds, similar experiences, similar successes, and similar training. That is what they know best: metallurgy. Any time we put five metallurgists on a team, we will typically have a metallurgical solution.

The same goes for any expertise in any discipline. This is the danger of not having technical diversity on a team and of letting an expert lead a team on an event in which the team members are the experts. Our greatest intellectual strengths represent liabilities when they lead us to miss something that we might have otherwise noticed—they create *blind spots*.[1]

The end to the story above is that eventually an engineer of a different discipline was assigned as the leader of the team. The new team had metallurgists as well as mechanical and process engineers. The result of the thorough Root Cause Analysis (RCA) was that the tubes that were rupturing were in a specific location of the boiler that was below the dew point for sulfuric acid. Therefore, the tubes were corroding due to the environment. The solution: return to the base metals and move the tubes 18 inches forward (outside of the brick wall) where the temperature was within acceptable limits.

When team leaders are NOT experts, they can ask any question they wish of the team members who should be the experts. However, this luxury is not afforded to experts who lead RCA teams because their team members generally perceive them as all-knowing. Therefore, they cannot ask the seemingly obvious or *stupid* question. While this seems a trivial point, it can, in fact, be a major barrier to success.

---

[1] Van Hecke, Madeleine L. 2007. *Blind Spots: Why Smart People Do Dumb Things.* New York: Prometheus Books, p. 22.

WHAT IS A TEAM?

A team is a small number of people with complementary skills who are committed to a common purpose, performance goals, and approach for which they hold themselves mutually accountable.[2]

A team is different than a group. A group can give the appearance of a team; however, the members act individually rather than in unison with others.

Let's explore the following key elements of an ideal RCA team structure:

1. Team Member Roles and Responsibilities
2. Principal Analyst (PA) Characteristics
3. The Challenges of RCA Facilitation
4. Promote Listening Skills
5. Team Codes of Conduct
6. Team Charter
7. Team Critical Success Factors (CSFs)

TEAM MEMBER ROLES AND RESPONSIBILITIES

Many views about ideal team size exist. The situation that created the team will generally determine how many members are appropriate. However, from an average standpoint for RCA, it has been our experience that between three and five core team members is ideal and beyond ten is too many. Having too many people on a team can force the goals to be prolonged due to the dragging on of too many opinions.

Who are the core members of an RCA team? They are as follows:

1. The PA
2. The Experts
3. Vendors
4. Critics

The Principal Analyst (PA)

Each RCA team needs a leader. This is the person who will ultimately be held accountable by management for results. They are the

---

[2] Katzenbach, Jon R. and Smith, Douglas K. 1994. *The Wisdom of Teams*. Boston, MA: Harvard Business School Press.

people who will drive success and accept nothing less. It is their desire that will either make or break the team. The PA should also be a facilitator, not a participator. This is a very important distinction because the technical experts who lead teams tend to participate instead of facilitating. The PA, as a facilitator only, recognizes that the answers are within the team members, and it is the PA's job to extract those answers in a disciplined manner by adhering to the PROACT® methodology.

This person is responsible for the administration of the team efforts, the facilitation of the team members according to the PROACT® philosophy, and the communication of goals and objectives to management oversight personnel.

The Experts

The experts are basically the core make-up of the team. These are the individuals the PA will facilitate. They are the *nuts-and-bolts* experts on the issue being analyzed. These individuals will be chosen based on their backgrounds in relation to the issue being analyzed. For instance, if we are analyzing an equipment breakdown in a plant, we may choose to have operations, maintenance, and engineering personnel represented on the team. If we are exploring an undesirable outcome in a hospital setting, we may wish to have doctors, nurses, lab personnel, and quality/risk management personnel on the team. In order to develop accurate hypotheses, experts are absolutely necessary on the team. Experts will aid the team in generating hypotheses and verifying them in the field.

Vendors

Vendors are an excellent source of information about their products. However, in our opinion, they should not lead an analysis when their products are involved in an event. Under such circumstances, we want the conclusions drawn by the team to be unbiased so that they have credibility. It is often very difficult for a vendor to be unbiased about how its product performed in the field. For this reason, we suggest that vendors participate on the team but not lead the team.

Vendors are great sources of information for generating hypotheses about how their products could not perform to expectations. However, they should not be permitted to prove or disprove their own hypotheses. We often see situations where the vendor will blame the way in which the product was handled or maintained as the cause of its nonperformance. It always seems to be something the customer did rather than a flaw in the product itself. We are not saying that the customer is always right, but from an unbiased standpoint, we must explore both possibilities: that the product has a problem as well as that the customer could have done something wrong to the product. Remember, facts lead such analyses, not assumptions!

Critics

We have never come across a situation in our careers where we had difficulty in locating critics. Every critic knows who he or she is in the organization. However, sometimes critics get a bad reputation just because they are curious. Critics are typically people who do not see the world the way that everyone else does. They are really the "devil's advocates." They will force the team to see the other side of the tracks and find holes in logic by asking persistent questions. They are often viewed as uncooperative and not team players. But they are a necessity to a team.

Critics come in two forms: (1) constructive and (2) destructive. Constructive critics are essential to success and are naturally inquisitive individuals who take nothing (or very little) at face value. Destructive critics stifle team progress and are more interested in overtime and donuts versus successfully accomplishing the Team Charter.

PRINCIPAL ANALYST CHARACTERISTICS

The PA typically has some tough challenges with managing an effective RCA team. If RCA is not part of the culture, PAs are going against the grain of the organization. This can be very difficult to deal with if PAs are people who have difficulty in dealing with barriers to success. Over the years, we have noted the personality traits that make certain PAs stars whereas others have not progressed. Following are the key traits that our most successful analysts portray:

1. Unbiased—nothing to lose or gain by the analysis outcome
2. Persistent—goes through hurdles, does not retreat
3. Organized—manages all steps of the investigative process
4. Diplomatic—knows how to deal with sensitive issues uncovered

THE CHALLENGES OF RCA FACILITATION

Those who have facilitated any type of team can surely appreciate the need to possess the characteristics described above. You can also appreciate the experience that such tasks provide in dealing with human beings. Next, we explore common challenges faced when facilitating a typical RCA team. We suspect that many of these will sound very familiar:

1. Bypassing the RCA Discipline and Going Straight to a Solution
2. Floundering of Team Members
3. Acceptance of Opinions as Facts
4. Dominating Team Members
5. Reluctant Team Members
6. Going Off on Tangents
7. Arguing among Team Members

A seasoned PA will stick to the RCA script (discipline) and not let these challenges get in the way of an effective analysis. They will know how to handle these situations (which will occur) while maintaining the synergy of the team.

PROMOTE LISTENING SKILLS

Obviously, many of the team dynamics issues we are discussing are not just pertinent to RCA but to any team. While the concept of listening seems simplistic, most of us are not adept at its use.

Many of us often state we are not good at remembering names. If we look back at a major cause of this, it is because we never actually listen to people when they introduce themselves to us for the first time. Most of the time when someone introduces himself or herself to us, we are more preoccupied with preparing our response than listening to what the person is saying.

Next time you meet someone, concentrate on listening to the person's introduction and take an imaginary *snapshot* of the person's face with your eyes. You will be amazed at how that impression will log into your long-term memory and pop up the next time you see that person.

The following are listening techniques that may be helpful when organizing RCA teams:

1. One Person Speaks at a Time
2. Don't Interrupt
3. React to Ideas, Not People
4. Separate Facts from Conventional Wisdom

TEAM CODES OF CONDUCT

Codes of conduct are merely sets of guidelines by which a team agrees to operate. Such guidelines are designed to enhance the productivity of team meetings. Following are a few examples of common sense codes of conduct:

1. All members will be on time for scheduled meetings.
2. All meetings will have an agenda that will be followed.
3. Everyone's ideas will be heard.
4. Only one person speaks at a time.
5. "3 Knock" rule will apply—This is where a person politely knocks on the table to provide an audio indicator that the speaker is going off track of the agenda topic being discussed.
6. "Holding Area"—This is a place on the easel pad where topics are placed for consideration on the next meeting agenda because it is not an appropriate topic for the meeting at hand.

This is just a sampling to give an idea as to what team meeting guidelines can be like.

TEAM CHARTER/MISSION

The team's charter (or sometimes referred to as a Mission) is a one-paragraph statement delineating why the team was formed in the first place. This statement serves as the focal point for the team. Such a statement should be agreed upon not only by the team but also by the managers overseeing the team's activities. This will align everyone's expectations as to the team's direction and expected results.

The following is a sample Team Charter reflecting a team that was organized to analyze a mechanical failure:

> To identify the root causes of the ongoing motor failures, occurring on pump CP-220, which includes identifying deficiencies in, or lack of, organizational systems. Appropriate recommendations for root causes will be communicated to management for rapid resolution.

TEAM CRITICAL SUCCESS FACTORS (CSFs)

CSFs are guidelines by which we will know that we are successful. I have heard CSFs also referred to as Key Performance Indicators (KPIs) and other nomenclature. Regardless of what we call them, we should set some parameters that will define the success of the RCA team's efforts. This should not be an effort in futility in listing a hundred different items. We recommend that no more than eight should be designated per analysis. Experience supports that typically many are used repeatedly on various RCA teams. Following are a few samples of CSFs:

1. A disciplined RCA approach will be utilized and adhered to.
2. A cross-functional section of site personnel/experts will participate in the analysis.
3. All analysis hypotheses will be verified or disproven with factual data.
4. Management agrees to fairly evaluate the analysis team's findings and recommendations upon completion of the RCA.
5. No one will be disciplined for honest mistakes.
6. A measurement process will be used to track the progress of implemented recommendations.

4

ANALYZING THE DATA
Introducing the PROACT® Logic Tree

No matter what methods to conduct Root Cause Analysis (RCA) are out in the marketplace, they all use either a categorical or cause-and-effect approach to determining causation. The various RCA methods in the marketplace may vary in presentation, but the legitimate ones are merely different in the way in which they graphically represent the determination of causation. Everyone will have their favorite RCA *tool*, which is fine, if they are using it properly and it is producing positive results.

CATEGORICAL VERSUS CAUSE-AND-EFFECT RCA TOOLS

Let's start with exploring the technical aspects of some of the more common RCA tools in the marketplace and contrasting them to each other. We will speak in generalities about these tools, as there is wide variability in how they are applied. Let's explore the following common RCA tools used in the marketplace today:

1. The 5 Whys
2. The Fishbone Diagram
3. The Regulatory Forms
4. The Logic Tree

ANALYTICAL TOOLS REVIEW

The goal of this description is not to teach you how to use these tools properly but to demonstrate how they can lack breadth and depth of approach and therefore affect the comprehensiveness of the analysis outcomes. **Any analytical tool is only as good as its user.**

Used properly and comprehensively, any of these tools can produce good, yet variable results. However, experience shows the attractiveness of these tools is their drawback as well.

The 5-Whys

Let's start with the 5 Whys. While there are varying forms of this simplistic approach, the most common understanding is that the analyst is to ask himself or herself the question "WHY?" five times and this will uncover *the* root cause. The form of this approach may be as shown in Figure 4.1.

There is a reason we do not hear about National Transportation Safety Board (NTSB) investigators using the 5 Whys approach during their investigations.

Here are how we see the pros and cons of the traditional 5 Whys approach applied to a serious event:

Pros
The attractive aspects of this approach are:

1. It's quick. Usually takes under an hour (if not minutes).
2. It's not resource intensive. Typically conducted by individuals as opposed to teams.
3. It concludes with one (1) root cause.
4. For the above reasons, it's inexpensive.

Figure 4.1 The 5 Whys analytical tool.

Cons

The main drawbacks with this concept, as a full-blown RCA tool for serious events are that:

1. Failure does not always occur in a linear pattern. Multiple factors combine laterally (parallel) to allow the undesirable outcomes to occur.
2. There is almost never a single root cause, and this is a misleading aspect of this approach as well.
3. People tend to use this tool as individuals and not in a team.
4. Hypotheses and conclusions are rarely backed up evidence.

The original intent of the development of the 5 Whys was for use by individuals working on the Toyota assembly lines. When faced with an undesirable outcome at the line level, individuals were encouraged to think deeper than they normally would to explore possible contributing factors to the outcome. If this did not resolve the issue, it would be passed on to a team that would investigate the issue using more comprehensive tools.

The Traditional Fishbone Diagram

The fishbone diagram is the second popular analytical Quality tool on the market. This approach gets its name from its form, which is in the shape of a fish (Figure 4.2). The spine of the fish represents the sequence of events leading to the undesirable outcome. The fish bones themselves represent cause categories that should be evaluated as having been potential contributors to the sequence of events. These categories change from user to user. The more popular cause category sets tend to be

The 6 M's: Management, Man, Method, Machine, Material, Measurement
The 4 P's: Place, Procedure, People, Policies
The 4 S's: Surroundings, Suppliers, Systems, Skills

The fishbone diagram is a tool often used in conjunction with brainstorming. Team members decide on the cause category set to be used and continue to ask what factors within the category caused the event to occur. Once these factors are identified and consensus is attained, attention is focused on solutions.

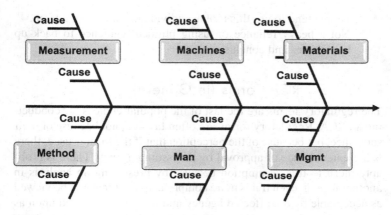

Figure 4.2 Sample fishbone diagram.

Here are how we see the pros and cons of the traditional fishbone approach applied to a serious event:

Pros
The attractive aspects of this approach are:

1. It's traditionally team based.
2. It's easy to set-up based on a selection of cause category sets (fishbones).
3. It's more comprehensive than traditional 5 Whys.
4. It's a simple approach to facilitate.

Cons
The main drawbacks with this approach for serious events:

1. This is a brainstorming technique at its core. After "Cause" categories are selected, the team brainstorms as to what could have occurred in each cause category. Thus, this is not a cause-and-effect-based approach.
2. If a less-than-adequate cause category set is chosen, full categories of potential causes/contributing factors could be missed.
3. This tool is less likely to depend on evidence to support hypotheses and more likely to let hearsay fly as fact.

4. People tend to use this tool as individuals and not in a team.
5. Not a heavy reliance on using physical evidence to back up hypotheses and conclusions.

Regulatory RCA Forms (in General)

The regulatory forms are the last of the popular choices for conducting an RCA. Regulatory forms are often favored, not because of their substance, but because of the perception that if the form is used, there is a greater chance of approval by the issuing agency. There is probably merit to this assumption because by presenting an analysis in another format, even if it is more comprehensive, it may not be viewed as acceptable by the affected agency and it would be looked upon as more work on the agency's part to make it fit their mold (their system).
Many regulatory forms have similar formats.

1. The first portion of the form normally deals with outcomes data. This is information related to the undesirable outcome or the consequences of the event.
2. The second portion of the form deals with determining causation. Most of the time this means that a series of questions (the same questions no matter the incident—one-size-fits-all type of approach) are asked or a list of cause categories are provided and the analyst is expected to brainstorm within those categories. This is very similar to the fishbone approach described earlier.
3. The third portion of the RCA Regulatory Form is often the Corrective Action portion of the form. After causation is determined, solutions are developed to overcome the identified causes within the categories.

Here are how we see the pros and cons of the use of Regulatory RCA Forms as a valid approach to conduct a serious event:

Pros
The attractive aspects of these forms:

1. They are easy to fill in the blanks.
2. They are more likely to be accepted (compliant) by the issuing agency in their own format.
3. One size fits all circumstances (mentality).

Cons

The main drawbacks with this approach for serious events:

1. Unfortunately, the characteristics of each undesirable outcome tend to be unique, and therefore, "one size fits all" solutions are typically not comprehensive enough to truly understand what happened in a bad outcome.
2. Leadership often believes that compliance equates profitability and safety.
3. Fosters a checklist mentality for conducting an RCA, which is dangerous when considering the risk of recurrence.
4. "RCA" is deemed a perfunctory task, RCA-by-the-numbers. This mentality will likely result in an increased risk of recurrence.
5. Not a heavy reliance on using physical evidence to back up hypotheses and conclusions.

Regulatory forms are a necessary evil when the success of the RCA effort is measured based on compliance. However, when measuring success based on a bottom-line metric, they often fall short.

For instance, we may be compliant, but does "being compliant" mean

1. There has been a decrease in the undesirable behaviors that led to the bad outcome?
2. There has been a decrease in affected operations and maintenance expenses?
3. There has been an increase in throughput?
4. There has been an increase in the Quality of the product and therefore value to the customer?
5. There has been an increase in client satisfaction?
6. There has been an increase in profitability?

When measuring the success of an RCA, we want to ensure that the metrics that measure our success reflect an improvement to the overall organization's goals. While compliance is certainly going to be a primary goal, we should get a more significant bottom-line benefit from being compliant. Identification of that benefit is a key in the success of our RCA effort.

The PROACT® Logic Tree

The PROACT® Logic Tree is representative of a tool specifically designed for use within RCA. The Logic Tree is an expression of cause-and-effect relationships that queued up in a particular sequence to cause an undesirable outcome to occur. These cause-and-effect relationships are validated with hard evidence as opposed to hearsay. The data leads the analysis—not the loudest expert in the room. The strength of the tool is such that it can be, and is, used in court to represent solid cases.

This chapter is about the construction of a Logic Tree during an RCA using the PROACT® methodology rule sets. We will elaborate on the details of the logic and demonstrate that the use of a comprehensive cause-and-effect tool like the Logic Tree will identify causes that normally would not be picked up in a categorical RCA approach.

Before we delve into the details of how to construct a Logic Tree, here are how we see the pros and cons of using such a comprehensive tool (and why we are dedicating an entire chapter to it):

Pros
The attractiveness of using a Logic Tree to conduct an RCA on a serious event:

1. It's very comprehensive when applied properly.
2. It's capable of diving deep past the decision maker and into deficiencies in organizational systems.
3. It REQUIRES hard evidence to validate hypotheses, root causes, and contributing factors.
4. It links the social sciences (why good people often make poor decisions) with the physical sciences (why components break).
5. It may take more time and resources, but it also is more likely to ensure we are not doing another RCA on the same issue again…because it was done right the first time!

Cons
The main drawbacks with this approach for serious events:

1. It is time consuming, primarily due to the efforts to gather evidence in a disciplined manner (as described in Chapter 2—PRESERVE)

2. It is resource consuming, primarily as it requires assignment of the proper subject matter experts (SMEs) to participate on the team (as described in Chapter 3—ORGANIZING THE TEAM)

3. It requires progressive leadership to demand and accept that flawed organizational systems contribute to poor decision-making in the field.

Comprehensive cause-and-effect tools allow analysts the opportunity to directly correlate deficient systems to poor decision-making to undesirable outcomes using hard evidence.

When using categorical approaches where we brainstorm about what we *think* happened within a cause category, usually we cannot directly correlate that cause to a poor decision that led to a bad outcome.

When properly using a cause-and-effect tool like the Logic Tree, it will drive us to identify the cause categories that played a role in the bad outcome, rather than our having to guess, when using a categorical RCA tool.

THE GERMINATION OF A FAILURE

Before getting into a detailed discussion about how to graphically express the sequence of events leading to an undesirable outcome, let's first briefly discuss the origin of a failure and the path it takes to the point that we have to do something about it.

Figure 4.3 expresses the origin of failure as coming from deficient organizational systems. Such systems include policies, procedures, training systems, and purchasing systems. Organizational systems are put in place to provide users of the systems information to make better decisions. When such systems are obsolete, inadequate, or nonexistent, we increase the opportunity for Human Error in decision-making and therefore undesirable outcomes.

As bad information is fed to an individual, the person must internalize this information along with his or her own training, experience, education, and past successes and determine what actions are appropriate. A poor decision will result in the triggering of a series of observable cause-and-effect relationships that, left unchecked, will eventually lead to an undesirable outcome that will have to be addressed whether we like it or not.

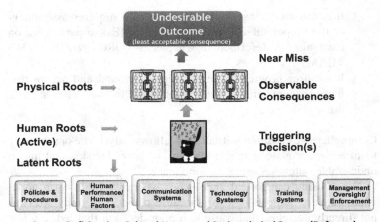

Figure 4.3   Germination of a failure.

In summary, these organizational systems represent the cause categories we have been describing. Specific causes identified within these categories are referred to as Latent Root (LR) Causes. As these deficient systems feed poor information to an individual, the individual is more apt to make a poor decision. We will refer to this decision error as a Human Root (HR) Cause. When humans take inappropriate actions, they trigger physical consequences to occur. We will refer to the initial physical consequences as Physical Root (PR) Causes. This root system will be described in detail later in this chapter. Understanding the germination of a failure is important before we attempt to try and graphically express it.

CONSTRUCTING A LOGIC TREE

Let's move on and delve into the details of constructing the PROACT® RCA tool of choice called a "Logic Tree." This is our means of organizing all the data collected thus far and putting it into an understandable and logical format for comprehension. This is different than the traditional logic diagram and a traditional fault tree. A logic diagram is typically a decision flowchart that will ask a question and, depending on the answer, will direct the user to a predetermined path of logic. Logic diagrams are popular in situations where the logic of a system has been laid out to aid in human decision-making. For instance, an operator in a nuclear power generation facility might use

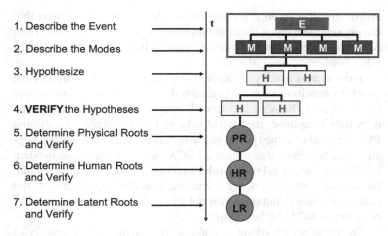

1. Describe the Event
2. Describe the Modes
3. Hypothesize
4. **VERIFY** the Hypotheses
5. Determine Physical Roots and Verify
6. Determine Human Roots and Verify
7. Determine Latent Roots and Verify

Figure 4.4 Logic Tree architecture.

such a logic diagram when an abnormal situation arises on the control panel and a quick response is required. A 911 operator might refer to a logic diagram under certain circumstances and ask the caller a series of questions. Based on the caller's answers, the string of questioning would change.

A fault tree is traditionally a totally probabilistic tool that utilizes a graphical tree concept that starts with a hypothetical event. For instance, we may be interested in how an event could occur so we would deduce the possibilities on the next lower level.

A Logic Tree is a combination of both of the above tools. The answer to certain questions will lead the user to the next lower level. However, the event and its surrounding modes (manifestations) will be factual versus hypothetical. Figure 4.4 shows the basic Logic Tree architecture. We will begin to dissect this architecture to gain a full understanding of each of its components in order to obtain a full understanding of its power.

THE EVENT

Following is a brief description of the undesirable outcome being analyzed. This is an extremely important block because it sets the stage for the remainder of the analysis. **THE EVENT MUST BE A FACT**. It cannot be an assumption. From an equipment standpoint, the event is typically the loss of function of a piece of equipment and/or process. From a production standpoint, it is the reason that the organization cares about the undesirable outcome. Under certain conditions, we will accept such an undesirable outcome, whereas in other conditions, we will not.

The *Event* is usually ill-defined, and there is no standard against which to benchmark, as no common definition exists. Many people believe they do RCA on incidents. However, if they look back on the ones they have done, they will likely find they probably were doing RCA because of some type of negative consequence. It is usually negative consequences that trigger an RCA, not necessarily the incident itself. Think about it. I may think I am doing an RCA because a pump failed, but I am really doing it because it stopped production. If the same pump failed and there was not a negative consequence, would I be doing an RCA on the failure?

The point we are trying to make is the magnitude and severity of the negative consequences will usually dictate whether an RCA will be commissioned and the depth and breadth of the analysis to be conducted.

When we are in a business environment that is not sold out (meaning we cannot sell all we can make), we are more tolerant of equipment failures that restrict capacity because we do not need the capacity anyway.

However, when sales pick up and the additional capacity is needed, we cannot tolerate such stoppages and rate restrictions. In the non-sold-out state, the event may be accepted. In the sold-out state, it is not accepted. This is what we mean by the event being defined as "the reason we care." We only care here because we could have sold the product for a profit.

THE MODE(S)

The modes are a further description of how the event HAS occurred in the past. REMEMBER, **THE EVENT AND MODE LEVELS MUST BE FACTS**. This is what separates the Logic Tree from a fault tree. It is a deduction from the event block and seeks to break down the bigger picture into smaller, more manageable blocks. Modes are typically easier to delineate when analyzing chronic events. Let's say here that we have a process that continually upsets. We lose production capacity for various reasons (modes).

In this case, the process *has been upset in the past* due to motor failure, pump failure, fan failure, and shaft failure. These modes represent individual occurrences. This does not mean that they do not have

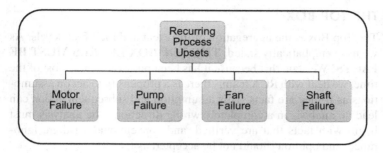

Figure 4.5   Top Box example of chronic event.

common causes, but their occurrences surfaced separately. Essentially the modes are answering the question, "How has the event occurred in the past?" In this case, the Top Box may look like Figure 4.5.

When dealing with sporadic events (one-time occurrences), we do not have the luxury of repetition so we must rely on the facts at the scene. The modes will represent the manifestation of the failure. The mode will be what triggered the negative consequence to occur. Sporadic failures may have fewer modes than chronic failures. This is because chronic failures represent more than one occurrence. It is not uncommon with a sporadic event to have a single mode.

Notice in this case that we included *inadequate response* as a mode. Why? Remember we said earlier that a mode was something that triggered the event to occur. This is also true of when we are looking at what actions after the incident occurred made it worse. Did our response increase the magnitude and severity of the consequences (the event)? By adding this mode, we will be seeking to identify our response to the incident and uncover when defensive systems were in place at that time and whether they were appropriate.

If they existed and were appropriate, did we follow them? If we did not follow them, why not? We can see from this line of questioning we will uncover the system flaws in our response. This way we can implement countermeasures to fill the cracks in our response plans in the future.

In our example, the existence of a failed pipe wall along with an improper response caused the initial limited effect of the failure to spread to the entire facility. In this scenario, a 600 lb steam line ruptured, and because the steam lines were not properly contained in a timely manner, all boilers were drained, shutting down the entire facility.

THE TOP BOX

The Top Box is the aggregation of the event and the Mode levels. As we have emphatically stated, **THE TOP BOX LEVELS MUST BE FACTS!** We state this because it has been our experience most of the time, we deal with RCA teams, there is a propensity to act on assumptions as if they are facts. This assumption and subsequent action can lead an analysis in a completely wrong direction. The analysis must begin with facts that are verified, and conventional wisdom, ignorance, and opinion should not be accepted as fact.

THE HYPOTHESES

As we learned in school at an early age, a hypothesis is merely an *educated guess*. Without making it any more complex than that, hypotheses are responses to the "How could ..." questions described previously.

This is the point at which the facts end, and we must hypothesize. So, we simply ask the question, "How could the preceding mode have occurred?" The answers we seek should be as broad and all-inclusive as possible. As we will show in the remainder of this chapter, this is contrary to normal problem-solving thought processes.

Let's take a few minutes here and discuss the nuances between asking the question "why" as opposed to "how could." Several self-proclaimed RCA techniques involve the use of asking the question *why*. Such tools include the 5 Whys and various types of Why Trees. Rather than get into the pros and cons of the approaches themselves, we will make one key distinction between them and PROACT®'s Logic Tree tool. When we ask the question *why*, we are connoting two things in our anticipated response: (1) that we seek a singular answer and (2) that we want an individual's opinion. From our standpoint, based on these premises, asking *why* encourages a narrow range of possibilities and allows assumption to potentially serve as fact. **If we are seeking someone's opinion without backing it up with evidence, it is an assumption!** This allows ignorance to creep into analyses and serve as fact.

On the flip side, what do we seek when asking *how could* something occur? This line of questioning promotes seeking all the possibilities instead of the most likely. Keep in mind the reason chronic events occur is because our conventional thinking has not been able to solve them in the past. Therefore, the true answers lie in something "unlikely" that will be captured by asking *how could* as opposed to *why*. Based on our responses to the *how could* questions, we will tap into our 5 P's data we collected earlier and use it to prove or disprove our hypotheses.

This distinction may seem like semantics, but it is a primary key to the success of any RCA. Only when we explore all the possibilities can we be assured that we have captured all the culprits. In PROACT® RCA what we prove *not* to be true is just as important (if not more important) as what we prove to be true.

VERIFICATIONS OF HYPOTHESES

As mentioned previously, hypotheses that are accepted without validation are merely assumptions. This approach, though a prevalent problem-solving strategy, is no more than a *trial-and-error* approach. In other words, it appears to be this case, so we will spend money on this fix. When that does not work, we reiterate the process and spend money on the next likely cause. This is an exhaustive and expensive approach to problem-solving. Typically, brainstorming techniques such as fishbone, 5 Whys, and troubleshooting approaches do not require validation of hearsay with evidence. This can be dangerous and expensive. Dangerous because all the causes have not been identified or verified, leaving us open to the risk of recurrence. Expensive because we may keep spending money until something finally works.

In the PROACT® RCA methodology, all hypotheses must be supported with hard data. The initial data for this purpose was collected in our 5 P's effort in the categories of Parts, Position, People, Paper, and Paradigms. The 5 P's data will ultimately be used to validate hypotheses on the Logic Tree. While this is a vigorous approach, the same parallel is used for the police detective preparing for court. The detective seeks a solid case and so do we. A solid case is built on facts, not assumptions. Would we expect a detective to win a murder case based on the sole testimony of a convicted drug dealer? This is a weak case and not likely to be successful.

The Top Box is equated to the crime scene or the facts. When all we must deal with are the facts, we start to question *how could* these facts exist in this form. The answers to these questions represent our hypotheses. To a criminologist, they represent leads. Leads must be validated with evidence, and all they do is *lead* to asking another question and the process continues. Eventually we will uncover what we call physical causes and what the detectives call forensic evidence.

Just like in the television series CSI (Crime Scene Investigation), people who do laboratory forensic work deal with the "how's" or the physical evidence. Their role is not to determine the *whys*.

The *whys* are analogous to motive and opportunity in a criminal investigation. As we will discuss later in this chapter, PROACT® associates the terms Human and Latent Root Causes with motive and opportunity. Prosecutors must prove why the defendant chose to take the actions he or she did that triggered the physical evidence to occur and eventually commit a crime.

For the purposes of this discussion on verification of hypotheses, we will use the following definition of evidence:

> **Evidence:** Any data used to prove or disprove the validity of a hypothesis during an investigation and/or analysis.

The literal definition of evidence will mean different things to different people based on their occupations. The meaning of evidence in the eyes of the law may be different than evidence to a root cause analyst. We have defined evidence in the manner above because it is simple, to the point, and represents how we use the term in RCA.

Hard data for validation means eyewitness accounts, factual statistics, certified tests, inspections, online measurement data, and the like. **A hypothesis proven to be true with hard data becomes a fact**.

It is important to note the concept of Cognitive Dissonance related to our interpretation of evidence in an RCA. Cognitive Dissonance is a state of tension that occurs whenever a person holds two cognitions (ideas, beliefs, attitudes, opinions) that are so psychologically inconsistent, such as "smoking is a dumb thing to do because it could kill me" and "I smoke two packs a day."[1] In layman's terms, we often tend to justify poor decisions despite the evidence presented. Part of our investigation must be to understand why we overrode the convincing evidence.

Along these same lines, we must also understand the effect of Confirmation Bias. If new information is consistent with our beliefs, we think it is well founded and useful: "Just what I always said." But if the new information is dissonant, then we consider it biased or foolish: "What a dumb argument!"[2] When we are leading analyses,

---

[1] Tavris, Carol and Elliott Aronson. 2007. *Mistakes Were Made (But Not by Me)*. Orlando: Harcourt, Inc., p. 12.

[2] Tavris, Carol and Elliott Aronson. 2007. *Mistakes Were Made (But Not by Me)*. Orlando: Harcourt, Inc., p. 18.

we should be very cognizant of Confirmation Bias and not allow our personal biases to interpret the evidence at hand.

This is especially tempting when dealing with our legal system, and our safety analysis may enter a legal realm. In legal circles, polarization of evidence is more likely than in a safety investigation. In court, we have a plaintiff and a defendant. The plaintiff's charges outline the plaintiff's case and the defense must counter that position. There are two sides looking at the same set of facts. Each side will interpret the facts to its own benefit (Confirmation Bias). Evidence that is entered that agrees with our case, we readily accept. Evidence that contradicts our case, we seek to discredit.

In keeping with our "solid case" analogy, we must keep in mind that organization is a key to preparing our case. To that end, we should maintain a Verification Log on a continual basis to document our supporting data.

THE FACT LINE

The fact line starts below the *mode* level because above it are facts and below it are hypotheses. As hypotheses are proven to be true with hard data and become facts, the fact line moves down the length of the tree.

PHYSICAL ROOT CAUSES (PR)

The first root-level causes that are encountered through the reiterative process will be the PRs. **PRs are the tangible roots or component-level roots**. PRs are observable. In many cases, when undisciplined problem-solving methods are used, people will tend to stop at this level and call them "Root Causes." We do not subscribe to this type of thinking. In any event, all PR Causes must also endure validation to prove them as facts. **PRs are generally identifiable on the Logic Tree by the fact that they are usually the first perceptible consequences after a human decision error has been made**.

HUMAN ROOT CAUSES (HR)

HR Causes will almost always trigger a PR Cause to occur. **HR Causes are decision errors**. These are either errors of omission or commission. This means that either we decided not to do something we should have done or we did something we were not supposed to do. Examples of errors of omission might be that we were so inundated with reactive work, we purposely put needed inspection work on the back burner to handle the failures of the day. An error of commission might be that we aligned a piece of equipment improperly because we did not know how to do it correctly.

HR Causes are not intended to represent the vaguely used term of Human Error. We use the HR only to represent a human decision that triggered a series of physical consequences to occur. In the end, this series of physical consequences ultimately resulted in an undesirable outcome. **Ending an analysis with a conclusion of "Human Error" is a cop out**. It is vague and usually indicates that it is not known why the incident occurred.

Oftentimes, we as the public are told that airplane accident investigations result in pilot error. This should be offensive to the general public because that pilot's life depended on the decision that they made, and they knew that. Therefore, we can reasonably conclude that he was making the best decisions he could at the time. What does that tell us? It can tell us many things, some of which include (1) the pilot was not trained properly for the situation he or she encountered, (2) the procedure the pilot did follow was inadequate for some reason, (3) the pilot did not follow the appropriate procedures (in which we would have to ask *why*), or the pilot was provided poor information from either the instrumentation or air traffic control. To simply say Human Error does not describe what happened.

PROACT® seeks to uncover the reasons people thought they were making the right decision at the time they made the decision. We refer to this basis of the decision as the LRs. These are the traps that result in poor decisions being made.

The frame of reference for understanding people's behavior and judging whether it made sense is in their own normal work context, the context in which they were embedded. This is the point of view from where decision assessments are sensible, normal, daily, unremarkable, and expected. The challenge, if we really want to know whether people anticipated risks correctly, is to see the world through their eyes, without knowledge of outcome, without knowing exactly which piece of data will turn out to be critical afterward.[3]

The point we wanted to make here is that by understanding the conditions that increase the risk of Human Error in decision-making,

---

[3] Dekker, Sidney. 2007. *Just Culture: Balancing Safety and Accountability.* Hampshire, England: Ashgate Publishing, p. 72.

we can implement proactive changes to reduce the risk. We are not perfect beings, so we will never eliminate Human Error in decision-making. The misattribution of errors is one reason we fail to learn from our mistakes: we haven't understood their root causes.[4] But that does not mean we cannot strive for such perfection, as success will be achieved during the journey.

While the questioning process thus far has been consistent with asking *how could*, at the HR level (decision error), we want to switch the questioning to "why?" When dealing in the physical and process areas, we cannot ask equipment *why* it failed. Only at the HR level do we encounter a person's involvement. When we get to this level, we are not interested in *whodunit* but rather why they made the decision that they did at the time they did. Understanding the rationale behind decisions that result in error is the key to conducting true RCA. Anyone who stops an RCA at the Human Level and disciplines an identified person or group is participating in a witch-hunt. Witch-hunts were discussed in the Preserving Failure Data section and proved to be non-value added, as the true roots cannot be attained in this manner. This is because if we search for a scapegoat, no one else will participate in the analysis for fear of repercussions. When we cannot find out why people make the decisions they do, we cannot permanently solve the issue at hand. Therefore, we cannot eliminate its risk of recurrence.

LATENT ROOT CAUSES (LR)

LR Causes are the organizational systems that people utilize to make decisions. When such systems are flawed, they result in decision errors. The term "Latent"[5] is defined as

> **Latent:** Those adverse consequences that may lie dormant within the system for a long time, only becoming evident when they combine with other factors to breach the system's defenses.

When we use the term organizational or management systems, we are referring to the *rules and laws* that govern a facility. Examples of organizational systems might include policies, operating procedures, maintenance procedures, purchasing practices, stores and inventory

---

[4] Hallinan, Joseph T. 2009. *Why We Make Mistakes*. New York: The Doubleday Publishing Group, p. 189.

[5] Reason, James. 1990–1992. *Human Error*. Victoria: Cambridge University Press, p. 173.

practices, training systems, and Quality control mechanisms. These systems are all put in place to help people make better decisions. When a system is inadequate or obsolete, people end up making decision errors based on flawed information. These are the true root causes of undesirable events. We have now defined the most relevant terms associated with the construction of a Logic Tree. Now let's explore the physical building of the tree and the thought processes that go on in the human mind.

Experts who participate on such teams are generally well-educated individuals, well respected within the organization as problem solvers, and people who pay meticulous attention to detail. With all this said and done, using the Logic Tree format, an expert's thought process may look like an event block going straight to Cause block, because there was no thoughtful "analysis"...just gut instinct because they are "the expert."

This poses a potential hurdle to a team's success, because for the most part, the analysis portion is bypassed, and we go straight from problem definition to cause. It is the Principal Analyst's responsibility to funnel the expertise of the team in a constructive manner without alienating the team members. Such an RCA team will tend to go to the *micro* view and not the *macro* view. However, in order to understand exactly what is happening, we must step back and look at the big picture. In order to do this, we must derive exactly where our thought process originated from and search for assumptions in the logic.

A Logic Tree is merely a graphical expression of what a thought would look like if it were on paper. It is looking at how we think. Let's take a simple example of a pump of some type that is failing. We find that 80% of the time this pump is failing due to a bearing failure. This shall serve as the *mode* that we pursue first for demonstration purposes.

BREADTH AND ALL-INCLUSIVENESS

If we have a team of operations, maintenance, and technical members and ask them the question, "How could a bearing fail?" their answers would likely get into the nuts and bolts of such details as improper installation, design error, defective materials, too much or too little lubricant, misalignment, and the like. While these are all very valid, they jump into too much detail too fast. We want to use deductive logic in short leaps.

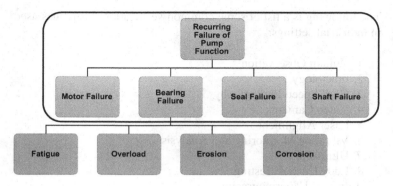

Figure 4.6 Broad and all-inclusive thinking.

In order to be *broad and all-inclusive* at each level, we want to identify all the possible hypotheses in the fewest blocks. To do this, we must imagine we are the part being analyzed. For instance, in the above example with the bearing, if we thought of ourselves as being the bearing, we would think, "How exactly did we fail?" **From a physical failure standpoint, the bearing would have to erode, corrode, overload, or fatigue. These are the only ways the bearing (or any component) can fail.** All the hypotheses developed earlier by the experts (the micro answers) would cause one or more of these failure mechanisms to occur (Figure 4.6).

From this point we would have a metallurgical review of the bearing conducted. If the results were to come back and state the bearing failed due to fatigue, then there are only certain conditions that can cause a fatigue failure to occur. The data or evidence leads us in the correct direction, not the team leader. This process is entirely data (fact) driven.

If we are broad and all-inclusive at each level of the Logic Tree and we verify each hypothesis with hard data, then the fact line drops until we have uncovered all the root causes. This is very similar in concept to many Quality initiatives. The more popular Quality initiatives focused on Quality of the entire manufacturing process instead of just checking Quality of the finished product (when it was too late).

VERIFICATION TECHNIQUES

While we used simple verification techniques in the previous example, there are thousands of ways in which to validate hypotheses. They are all, obviously, dependent on the nature of the hypothesis.

The following is a list of some common verification techniques used in industrial settings:

1. Human Observation
2. Fractology
3. High-Speed Photography
4. Video Cameras
5. Laser Alignment
6. Vibration Monitoring and Analysis
7. Ultrasonics
8. Eddy Current Testing
9. Infrared Thermography
10. Ferrography
11. Scanning Electron Microscopy
12. Metallurgical Analysis
13. Chemical Analysis
14. Statistical Analysis (correlation, regression, Weibull Analysis, etc.)
15. Operating Deflection (OD) Shapes
16. Finite Element Analysis (FEA) Modeling
17. Motor Circuit Analysis
18. Modal Analysis
19. Experimental Stress Analysis
20. Rotor Dynamics Analysis
21. Task Analysis

These are just a few techniques to give you a feel for the breadth of verification techniques that are available. There are literally thousands more. Each of these topics could be a text as well. Many texts are currently available to provide more in-depth knowledge on each of these techniques. However, the focus of this text is on the PROACT® RCA methodology. A good Principal Analyst does not necessarily have to be an expert in any or all of these techniques; rather, he or she should be resourceful enough to know when to use which technique and how to obtain the resources to complete the test. Principal Analysts should have a repository of resources they can tap into when the situation permits.

CONFIDENCE FACTORS

It has been our experience that the timelier and more pertinent the data that is collected about a specific event, the quicker the analysis is completed and the more accurate the results are. Conversely, the less data we have initially, the longer the analysis takes and the greater the risk of the wrong cause(s) being identified.

We utilize a *confidence factor* rating for each hypothesis to evaluate how confident we are with the validity of the test and the accuracy of the conclusion. The scale is basic and runs from 0 to 5. A "0" means without a doubt, with 100% certainty, based on the validated data collected the hypothesis is *NOT* true. On the flip side, a "5" means that based on the data collected and the tests performed, there is 100% certainty the hypothesis *IS* true. Between the "0" and the "5" are the shades of gray where the data used was not conclusive. This is not uncommon in situations where an RCA is commissioned weeks after the event occurred and little or no data from the scene was collected. Also, in catastrophic explosions, we have seen that uncertainty resides in the physical environment prior to the explosion. What formed the combustible environment? These are just a few circumstances in which absolute certainty cannot be attained. The confidence factor rating communicates this level of certainty and can guide corrective action decisions.

We use the rule of thumb that a confidence factor rating of "3" or higher is treated as if it did happen and we continue to pursue the logic leg. Any confidence factor rating of less than "3" we treat as a low probability of occurrence and feel it should not be pursued at this time. However, the only hypotheses that are crossed out on the Logic Tree are the ones that have a confidence factor rating of "0". A "1" cannot be crossed out because it still had a probability of occurring even if the probability was low.

THE TROUBLESHOOTING FLOW DIAGRAM

Once the Logic Tree is completed, it should serve as a troubleshooting flow diagram for the organization (Figures 4.7 and 4.8). Chances are the root causes identified in this RCA will affect the rest of the organization. Therefore, some recommendations will be implemented site-wide or corporation-wide. To optimize the use of a world-class RCA effort, the goal should be the development of a dynamic troubleshooting flow diagram repository. This will end up containing logic diagrams or knowledge management templates that capture the expertise of the organization's best problem solvers on paper. We referred to this as *corporate memory*.

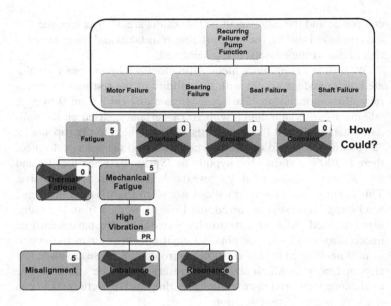

Figure 4.7   Sample Logic Tree (1).

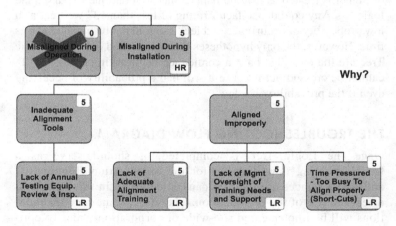

Figure 4.8   Sample Logic Tree (2).

Such logic diagrams can be stored on the company's intranet and be made available to all facilities that have similar operations and can learn from the work done at one site. These logic diagrams are complete with test procedures for each hypothesis. They are dynamic because where one RCA team may not have followed one particular hypothesis (because it was not true in their case), the hypothesis may be true in another case and the new RCA team can pick up from that point and explore the new logic path.

The goal of the organization should be to capture the intellectual capital of the workforce and make it available for all from which to learn. This optimizes the intellectual capital of the organization through RCA.

5

COMMUNICATING FINDINGS AND RECOMMENDATIONS

Let's assume at this point the complete Root Cause Analysis (RCA) process has been followed to the letter. We have determined our "Significant Few" failures to analyze. We have chosen a specific significant event and proceeded through the PROACT® process. An identified RCA team has undertaken an organized data collection effort. The team's charter and Critical Success Factors (CSFs) have been determined and a Principal Analyst (PA) has been named. A Logic Tree has been developed where all hypotheses have been either proven or disproved with hard data. Physical, Human, and Latent roots have been identified. Are we done?

Not quite! Success can be defined in many ways, but an RCA should not be deemed successful unless something has improved as a result of implementing recommendations from the RCA. **Merely conducting an excellent RCA does not produce results. As many of you can attest, getting something done about your findings can be the most difficult part of the analysis**. Oftentimes recommendations will fall on deaf ears, and the entire effort was a waste of your time and the company's money.

DEVELOPING THE RECOMMENDATIONS

Every corporation will have its own standards in how it wants recommendations to be written. It will be the RCA team's goal to abide by these internal standards while accomplishing the objectives of the RCA's Team Charter.

The core team members, at a predetermined location and time, should discuss recommendations. The entire meeting should be set aside to concentrate on recommendations alone. Remember our analogy of the detective throughout this text, always trying to build a solid case. This report and its recommendations represent our "day in court." In order to win the case, our recommendations must be solid and well thought out. But foremost, they must be accepted, implemented, and effective in order to be successful.

At this team meeting, the objective should be to gain team consensus on recommendations brought to the table. **Team consensus is not team agreement.** Team agreement means everyone gets what he or she wants. Team consensus means everyone can live with the content of the recommendations. Everyone did not get all of what they wanted, but they can live with it. Team agreement is rare.

The recommendations should be clear, concise, and understandable. Always have the objective in mind of eliminating or greatly reducing the impact of the cause when writing the recommendations. Every effort should be made to focus on the RCA. Sometimes we tend to have pet projects we attach to an RCA recommendation because it might have a better chance of being accepted. We liken this to riders on bills reviewed in Congress. They tend to bog down a good bill and threaten its passage in the long run. At the first sight of unnecessary recommendations, decision makers will begin to question the credibility of the entire RCA. When writing recommendations, stick to the issues at hand and focus on eliminating the risk of recurrence.

When the team develops recommendations, it is a good idea to present decision makers with multiple alternatives. Sometimes when we develop recommendations, they might be perceived as not meeting the predefined criteria given by management. If this is the case, then efforts should be made to have an alternative recommendation— a recommendation that clearly fits within the defined criteria. One thing we never want to happen is for an issue under which the presenters have some control to stall the management presentation. Absence of an acceptable recommendation is one such obstacle, and every effort should be made to gain closure of the RCA recommendations at this meeting.

DEVELOPING THE REPORT

The report represents the documentation of the "solid case" for court or, in our circumstances, the final management meeting. This should serve as a living document in that its greatest benefit will be that others learn from it to avoid the recurrence of similar events at other sites within the company or organization. To this end, the professionalism of the report should suit the nature of the event being analyzed. We like to use the adage, "If the event costs the corporation $5, then perform a $5 RCA. If it costs the organization $1,000,000, then perform a $1,000,000 type of RCA."

We should keep in mind that if RCAs are not prevalent in an organization, then the first RCA report usually sets the standard. We should be cognizant of this and take it into consideration when developing our reports. Let's assume at this point that we have analyzed a "Significant Few" event and it is costly to the organization. Our report will reflect that level or degree of importance.

The following table of contents will be our guide for the report:

1. The Executive Summary
    a. The Event Summary
    b. The Event Mechanism
    c. The PROACT® Methodology Description
    d. The Root Cause Action Matrix
2. The Technical Section
    a. The Identified Root Causes
    b. The Type of Root Causes
    c. The Responsibility of Executing Recommendations
    d. The Estimated Completion Dates
    e. The Detailed Plan to Execute Recommendations
3. Appendices
    a. Recognition of All Participants
    b. The 5 P's Data Collection Strategies
    c. The Team Charter
    d. The Team CSFs
    e. The Logic Tree
    f. The Verification Logs

Now let's review the significance and contribution of each element to the entire report and the overall RCA objectives.

THE EXECUTIVE SUMMARY

The Executive Summary is just that—a summary. It has been our experience that the typical decision makers at the upper levels of management are not nearly as concerned with the details of the RCA as they are with the results and credibility of the RCA. This section should serve as a synopsis of the entire RCA—a quick overview. This section is meant for managers and executives to review the analyzed event, the reason it occurred, what the team recommends ensuring it never happens again, and how much it will cost.

The Event Summary

The Event Summary is a description of what was observed from the point in time that the event occurred until the point in time that the event was isolated or contained. This can generally be thought of as a timeline description.

The Event Mechanism

The Event Mechanism is a description of the findings of the RCA. It is a summary of the errors that lead up to the point in time of the event occurrence. This is meant to give management a quick understanding of the chain of errors that were found to have caused the event in question.

The PROACT® Description

The PROACT® Description is a basic description of the PROACT® process for management to review. Sometimes management may not be aware of a formalized RCA process being used in the field. A basic description of such a disciplined and formal process generally adds credibility to the analysis and assures management that it was a professional effort.

The Root Cause Action Matrix

The Root Cause Action Matrix is a table outlining the results of the entire analysis. This table is a summation of identified causes, overview of proposed recommendations, person responsible for executing recommendations, and estimated completion date.

THE TECHNICAL SECTION

The Technical Section is where the details of all recommendations are located. This is where the technical staff may want to review the details of the analysis recommendations.

The Identified Root Cause(s)

The Identified Root Cause(s) will be delineated in this section as separate line items. All causes identified in the RCA that require countermeasures will be listed here.

The Type of Root Cause(s)

The Type of Root Cause(s) will be listed here to indicate their nature as being Physical, Human, or Latent Root Cause(s). It is important to note that only in cases of intent with malice should any indications be made as to identifying any individual or group. Even in such rare cases, it may not be prudent to specifically identify a person or group in the report because of liability concerns. Normally, no recommendations are required or necessary where a Human Root is identified. This is because if we address the Latent Root or the decision-making basis that led to the occurrence of the event, then we should subsequently change the behavior of the individual. For instance, if we have identified a Human Root as "Misalignment" of a shaft (no name necessary), then the actions to correct that situation might be to provide the individual the training and tools to align the shaft properly in the future. This countermeasure will address the concerns of the Human Root without making a specific Human Root recommendation or potentially giving the perception of blaming individuals or groups.

The Responsibility of Executing the Recommendation

This will also be listed to identify an individual or group that will be accountable for the successful implementation of the recommendation.

The Estimated Completion Date

The Estimated Completion Date will be listed to provide an estimated timeline for when each countermeasure will be completed, thus setting the anticipated timeline of returns on investment.

The Detailed Plan to Execute Recommendation

This section is generally viewed as an expansion of the Root Cause Action Matrix described previously. Here is where all the economic justifications, the plans to resource the project (if required), the funding allocations, etc., are located.

APPENDICES
Recognition of All Participants

Recognizing all participants is extremely important if our intent is to have team members participate on RCA teams in the future. It is suggested to note every person that inputs any information into the analysis in this section. All people tend to crave recognition for their successes.

The 5 P's Data Collection Strategies

These strategies should be placed as an addendum or appendix item to show the structured efforts to gain access to the necessary data to make the RCA successful.

The Team Charter

The Team Charter should also be placed in the report to show the team displayed structure and focus about their efforts.

The Team Critical Success Factors

Including the Team CSFs shows the teams had guiding principles and defined the parameters of success.

The Logic Tree

The Logic Tree is a necessary component of the report for obvious reasons. The Logic Tree will serve as a dynamic expert system (or troubleshooting flow diagram) for future analysts. This type of information will optimize the effectiveness of any corporate RCA effort by conveying valuable information to other sites with similar events.

The Verification Logs

The Verification Logs are the spine of the Logic Tree and a vital part of the report. This section will house all the supporting documentation (evidence) for hypothesis validation.

REPORT USE, DISTRIBUTION, AND ACCESS

The report will serve as a living document. If a corporation wishes to optimize the value of its intellectual capital using RCA, then the issuing of a formal professional report to other relevant parties is necessary. Serious consideration should be given to RCA report distributions. Analysts should review their findings and recommendations and identify others in their organization who may have similar operations and, therefore, similar problems. These identified individuals or groups should be put on a distribution list for the report so they are aware that this particular event has been successfully analyzed and recommendations have been identified to eliminate the risk of recurrence. This optimizes the use of the information derived from the RCA.

In our information era, instant access to such documents is a must. Most corporations have their own internal intranets. This provides an opportunity for the corporation to store these newly developed "dynamic expert systems" in an electronic format allowing instant access. Corporations should explore the feasibility of adding such information to their intranets and allowing all sites to access the information. Using RCA software like PROACT® will make all RCA information and dashboards more accessible to stakeholders.

Whether the information is in paper or electronic format, the ability to produce RCA documentation quickly could help some organizations from a legal standpoint. Whether it is a government regulatory agency, corporate lawyers, or insurance representatives, demonstrating that a disciplined RCA method was used to identify root causes can prevent some legal actions against the corporations as well as fines from being imposed due to noncompliance with regulations. Most regulatory agencies that require a form of RCA to be performed by organizations do not delineate the RCA method to be used but rather ensure that one can be demonstrated upon audit.

THE FINAL PRESENTATION

This is the PA's "final day in court." It is what the entire body of RCA work is all about. Throughout the entire analysis, the team should be focused on this meeting. We have used the analogy of the detective throughout this text. In the chapter "Preserving Failure Data" (Chapter 2), we described why a detective goes to the lengths that he or she does in order to collect, analyze, and document data. Our conclusion was that the detective knows he or she is going to court and the lawyers must present a solid case in order to obtain a conviction.

Our situation is not much different. Our court is a final management review group that will decide if our case is solid enough to approve the requested monies for implementing recommendations.

Realizing the importance of this meeting, we should prepare accordingly. Preparation involves the following steps:

1. Have the professionally prepared reports ready and accessible.
2. Strategize for the meeting by knowing your audience.
3. Have an agenda for the meeting.
4. Develop a clear and concise professional presentation.
5. Coordinate the media to use in the presentation.
6. Conduct "dry runs" of the final presentation.
7. Quantify the effectiveness of the meeting.
8. Prioritize recommendations based on impact and effort.
9. Determine "next step" strategy.

We will address each of these individually and in some depth to maximize the effectiveness of the presentation and ensure that we get what we want.

HAVE THE PROFESSIONALLY PREPARED REPORTS READY AND ACCESSIBLE

At this stage, the reports should be ready, in full color and bound. Have a report for each member of the review team as well as for each team member. Part of the report includes the Logic Tree development. The Logic Tree is the focal point of the entire RCA effort and should be graphically represented as such. The Logic Tree should be printed on blueprint-sized paper, in full color and laminated if possible. The Logic Tree should be proudly displayed on the wall in full view of the review committee. Keep in mind that this Logic Tree will likely serve as a source of pride for the management to show other divisions, departments, and corporations how progressive their area is in conducting RCA. It will truly serve as a trophy for the organization.

STRATEGIZE FOR THE MEETING BY KNOWING YOUR AUDIENCE

This is an integral step in determining the success of the RCA effort. Many people believe they can develop a top-notch presentation that will suit all audiences. This has not been our experience. All audiences are different and, therefore, have different expectations and needs.

Consider our courtroom scenario again. Lawyers are courtroom strategists. They will base their case on the make-up of the jury and the judge presiding. When the jury has been selected, the lawyers will determine their backgrounds, whether they are middle class or upper class, etc. What is the ratio of men to women? What is the ethnic make-up of the jury? What is the judge's track record on cases like this one? What were the bases of the previous cases on which the judge has ruled? Apply this to our scenario, and we begin to understand that learning about the people we must influence is a must.

In preparing for the final presentation, determine which attendees will be present. Then learn about their backgrounds. Are they technical people, financial people, or perhaps marketing and salespeople? This will be of great value because making a technical presentation to a financial group would risk the success of the meeting.

Next, we must determine what makes these people "tick." How are these people's incentives paid? Is it based on throughput, cost reduction, profitability, various ratios, or safety records? This becomes very important because when making our presentation, we must present the benefits of implementing the recommendations in units that appeal to the audience. For example, "If we are able to correct this startup procedure and provide the operators the appropriate training, based on past history, we will be able to increase throughput by 1%, which will equate to $5 million annually."

HAVE AN AGENDA FOR THE MEETING

No matter what type of presentation media you have, always have an agenda prepared for such a formal presentation. Management typically expects this formality, and it also shows organizational skills on the part of the team. Below is a sample agenda that we typically follow in our RCA presentations.

1. Review of the PROACT® RCA Process Steps
2. Summary of the Undesirable Event
3. Description of Error Chains Found in RCA
4. Logic Tree Review (Sample Path)
5. Root Cause Action Matrix Review
6. Recognition of Participants Involved

7. Q&A Session
8. Commitment to Action/What's Next?

Always follow the agenda; only divert when requested by the management team.

Notice the last item on the agenda is "Commitment to Action." This is a very important agenda item as sometimes we tend to leave such meetings with a feeling of emptiness and we turn to our partner and ask, "How do you think it went?" Until this point, we have done a great deal of work and we should not have to "wonder how it went." It is not impolite or too forward to ask at the conclusion of the meeting, "Where do we go from here?" **Even a decision to do nothing is a decision and you know where you stand. Never leave the final meeting wondering how it went.**

DEVELOP A CLEAR AND CONCISE PROFESSIONAL PRESENTATION

Research shows the average attention span of individuals in managerial positions is about 20–30 minutes. The presentation portion of the meeting should be designed to accommodate this time frame. We recommend the entire meeting last no more than 1 hour. The remaining time will be left to review recommendations and develop action plans.

The presentation should be molded around the agenda we developed earlier. Typical presentation software such as Microsoft's PowerPoint®[1] provides excellent graphic capabilities and easily allows the integration of words, digital images, and animation. Remember this is our chance to communicate our findings and recommendations. Therefore, we must be as professional as possible to get our ideas approved for implementation. The use of various forms of media during a presentation provides an interesting forum for the audience and aids in retention of the information by the audience.

There is a complete psychology behind how the human mind tends to react to various colors. This type of research should be considered during presentation development. The use of laptops, LCD projectors, and easel pads will help in providing an array of different media to enhance the presentation. Props such as failed parts or pictures from the scene can be used to pass around to the audience and enhance

---

[1] PowerPoint® is a registered trademark of Microsoft.

interest and retention. All of this increases the chances of acceptance of the recommendations.

Always dress the part for the presentation. Our rule of thumb has always been to dress one level above the audience. We do not want to appear too informal, but we also do not want to appear too over-dressed. The key is to make sure your appearance is professional. Remember, we perform a $5 failure analysis on a $5 failure. This presentation is intended for a "Significant Few" item, and the associated preparation should reflect its importance.

COORDINATE THE MEDIA TO USE IN THE PRESENTATION

As discussed earlier, many forms of media should be used to make the presentation. To that end, coordination of the use of these items should be worked out ahead of time to ensure proper "flow" of the presentation. This is very important as lack of such preparation could affect the results of the meeting and give the presentation a disconnected or unorganized appearance.

Assignment of tasks should be made prior to the final presentation. Such assignments may include one person to manipulate the computer while another presents, a person to hand out materials or props at the speaker's request, and a person who will provide verification data at the request of management. Such preparations and organization really shine during a presentation and it is apparent to the audience.

It is also very important to understand the layout logistics of the room in which you are presenting. Nothing is worse than showing up at a conference room and realizing your laptop does not work with the LCD projector. Then you spend valuable time fidgeting with it and trying to make it work in front of your audience. Some things to keep in mind to this end include the following:

1. Know how many will be in your audience and where they will be sitting.
2. Use name cards if you wish to place certain people in certain positions in the audience.
3. Ensure that everyone can see your presentation from where they are sitting.
4. Ensure that you have enough handout material (if applicable).
5. Ensure that your A/V equipment is fully functional prior to the meeting.

Like everything else about RCA, we must be proactive in our preparation for our final presentation. After all, if we do not do well in this presentation, our RCA will not be successful, and thus, we will not have improved the bottom line.

CONDUCT "DRY RUNS" OF THE FINAL PRESENTATION

The final presentation should not be the testing grounds for the presentation. No matter how prepared we are, we must display some modesty and realize there is a possibility that we may have holes in our presentation and our logic.

We are advocates of at least two "dry runs" of the presentation being conducted prior to the final one. **We also suggest that such dry runs be presented in front of the best and most constructive critics in your organization.** Such people will be happy to identify logic holes, thereby strengthening the logic of the tree. The time to find gaps in logic is prior to the final presentation, not during. Logic holes that are found during the final presentation will ultimately damage the credibility of the entire Logic Tree. This is a key step in preparation for the final presentation.

PRIORITIZE RECOMMENDATIONS BASED ON IMPACT AND EFFORT

Part of getting what we want from such a presentation involves presenting the information in a "digestible" format. For instance, if you have completed an RCA and have developed 19 recommendations, the next task is to get them completed. As we well know, if we put 19 recommendations on someone's desk, there is a reduced likelihood that any will get done. Therefore, we must present them in a "digestible" manner. We must present them in such a format that it does not appear to be as much as it really is. How do we accomplish this task?

We utilize what we call an impact–effort priority matrix. This is a simple three-by-three table with the X-axis indicating impact and the Y-axis indicating effort to complete.

Let's return to our previous scenario of having 19 recommendations. At this point, we can separate the recommendations over which we have direct control to execute and determine them to be high impact, low effort recommendations. Maybe we deem several other recommendations as requiring the approval of other departments;

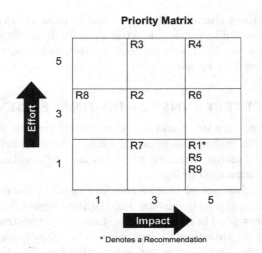

Figure 5.1 Completed impact–effort priority matrix.

therefore, they may be a little more difficult to implement. Finally, maybe we determine some recommendations require a shutdown to occur before the corrective action can be taken. Therefore, these recommendations are more difficult to implement. This is a subjective evaluation that breaks down the perception of too many recommendations into manageable and accomplishable tasks. A completed matrix may look like Figure 5.1.

DETERMINE NEXT STEP STRATEGY

The ultimate result we are looking for from this step (Communicate Findings and Recommendations) is a corrective action plan. This entire chapter dealt with selling the recommendations and gaining approvals to implement them. After the meeting, we should have recommendations that have been approved, individuals assigned to execute them, and timelines in which to have them completed. The next phase we will explore is the effectiveness of the implementation and overall impact on bottom-line performance.

6

TRACKING FOR
BOTTOM-LINE RESULTS

Consider what we have accomplished thus far in the PROACT® Root
Cause Analysis (RCA) process:

1. Developed a data preservation strategy
2. Organized an ideal RCA team
3. Utilized a disciplined method to draw accurate root causes
4. Prepared a formal RCA report and presentation for management
5. Defined which corrective actions will be implemented

This is an immense amount of work and an accomplishment in and of
itself. However, success is not defined as identifying root causes and
developing recommendations. Something must improve as a result of
implementing the recommendations!

Always keep in the back of our minds that we are continually selling our need to survive, whether it is in society or in our organization. We must be constantly proving why we are more valuable to
the facility than others. **Tracking for results becomes the measurement of our success in the RCA effort**. Therefore, since this reflects
our work, we should be diligent in measuring our progress because it
will be viewed as a report card of sorts. Once we establish successes,
we must exploit them by publicizing them for maximum personal and
organizational benefit. The more people who are aware and recognize
the success of our efforts, the more they will view us as people to
depend on in order to eliminate problems. This makes us a valuable
resource to the organization. Make note that if we are successful at
RCA, the reward should be we get to do it again. This will result in
the various departments or areas requesting the RCA service from us.
While this is a good indication, there can be drawbacks.

For instance, we have been trained to work on the "Significant
Few" events costing 80% of the organization's losses. Under the

described circumstances, we may have numerous people asking us to solve their smaller problems, which are not necessarily important to the organization. Therefore, when we decline, we may be viewed as not being a team player because we insisted on sticking to the "Significant Few" list (the 80/20 split). These are legitimate concerns we should address with our RCA Champions and Principal Analysts.

Let's pick up from the point where management has approved various recommendations of ours in our final meeting. Now what happens? We must consider each of the following steps:

1. Getting proactive work orders accomplished in a reactive environment
2. Sliding the proactive work scale
3. Developing tracking metric(s)
4. Exploiting successes
5. Creating a critical mass
6. Recognizing the lifecycle effects of RCA on the organization

GETTING PROACTIVE WORK ORDERS ACCOMPLISHED IN A REACTIVE ENVIRONMENT

Unless approved recommendations are implemented, we certainly cannot expect phenomenal results. Therefore, we must be diligent in our efforts to push the approved recommendations all the way through the system. One roadblock we have repeatedly run into is the fact that people generally perceive recommendations from RCAs as improvement work or proactive work. **In the midst of a reactive backlog of work orders, a proactive one does not stand a chance for implementation**.

Most Computerized Maintenance Management Systems (CMMSs), or their industry equivalents, possess a feature by which work orders are prioritized. Naturally, anyone who creates a work request thinks that their work is more important than anyone else's; therefore, they put the highest priority on the work request. The priority ranking system of any work order system goes something like this (or equivalent):

E = Emergency—Respond Immediately
1 = 24-Hour Response Required
2 = 48-Hour Response Required
3 = 1-Week Response Required

What normally happens with such prioritization systems is that many corrective work requests are entered as "E" or emergency events requiring the original schedule to be broken in order to accommodate them. Usually the preventive and predictive inspections are the first items to get removed from the schedule—the proactive work!

Given this scenario, what priority would a recommendation from an RCA have? Typically, a "4"! Such work is deemed as backburner work, and it can wait because the event is not occurring now. This is an endless cycle if the chain is not broken. This is like waiting to fix the hole in the roof until it rains.

We mentioned earlier that management systems must be put into place to support RCA efforts. This is one system that must be in place prior to even beginning RCA. If the recommendations are never going to be executed, then the RCA should never begin. Accommodations must be made in the work order system to give proactive work a fair chance of being accomplished against the reactive work. This will involve planners and schedulers to agree that a certain percentage of the maintenance resources must be allocated to executing the proactive work, no matter what. This is hard to do, both in theory and practice. But the fact of the matter is if we do not take measures to prevent the recurrence of undesirable events, we are acknowledging defeat against them and accepting reaction as the maintenance strategy. If we do not initially allocate some degree of resources to proactive work, we will always be stuck in a reactive cycle.

Keep in mind that "work orders" are synonymous with "repairs," not dealing with organizational system-related causes (Latent Root Causes). Therefore, most work order systems are not equipped to handle these "soft" root causes associated with people and systems.

The answer to the above paradox can be quite simple. We have seen companies simply identify a designation for proactive work and ensure that the planners and schedulers treat them as if they are an "E" ticket with the resources they have set aside to address such opportunities. Maybe it's a "P" for proactive work or a block of worker numbers. Whatever the case may be, consideration must be given to making sure that proactive work orders generated from RCAs are implemented in the field.

This needs to be a priority for the company and tracked as such. We need to track the amount of proactive work being done on a monthly basis. If the level of proactive work is insufficient, we need to make our plant RCA Champion aware so they can address the issues.

Most organizations do not like change. We are all in favor of improving things as long as we do not have to "change." Utilizing metrics to measure our level of proactive work will demonstrate how committed we are to improvement and defect elimination.

SLIDING THE PROACTIVE WORK SCALE

As we hear all the time, the most common objection to performing RCA in the field is that we do not have the time. When we look at this objection introspectively, we find we do not have the time because we are too busy reacting to failures and repairs. This truly is an oxymoron. RCA is designed to eliminate the need to react to unexpected failures. Managements must realize this and include RCA as part of the overall plant strategy.

One way we have seen this done is through an interactive board game developed originally within DuPont®[1] and now licensed through a company in Kingwood, TX, called The Manufacturing Game®.[2] Organizational development experts within DuPont® developed this game. It is an innovative way to involve all perspectives of a manufacturing plant. When we played The Manufacturing Game®, we found it to be an invaluable tool for demonstrating why a facility must allocate some initial resources to proactive work in order to remain competitive and in business. The Manufacturing Game® demonstrates why proactive activities are needed to eliminate the need to do work and RCA expresses how to actually do it.

Proaction and reaction should be inversely proportional. The more proactive tasks performed, the less reactive work there should be. Therefore, all the personnel we currently have conducting strictly reactive work will now have more time to face the challenges of proactive work. We have yet to see a facility that admittedly has all the resources it would like to have in order to conduct proactive tasks such as visual inspections, predictive maintenance, preventive maintenance, RCA, and lubrication. It does not have these resources now because they are in reactionary situations. As the level of proaction increases, the level of reaction will decrease. This is a point where we gain control of the operation and the operation does not control us! Research demonstrates that a reactive plant spends 25% more on routine maintenance than its counterparts in the proactive domain.

---

[1] DuPont is a registered trademark of the E. I. DuPont de Nemours & Co.
[2] ®1998 The Manufacturing Game.

It has also been proven there is a direct correlation between the amount of money we spend on maintenance and the losses associated with production disruptions. Some studies suggest that for every dollar that is spent on maintenance, there is a $4–$10 loss in production. This does not even address the safety and environmental issues linked to reactive work environments.

DEVELOPING TRACKING METRICS

Recognizing the inverse relationship between proaction and reaction, we must focus on how to measure the effects of implemented recommendations. This is generally not a complex task because typically there was an existing measurement system in place that identified a deficiency in the first place. By the time the RCA is completed, and the causes all identified, the metric to measure usually becomes obvious.

Let's review a few circumstances to determine appropriate metrics:

1. *Mechanical*—We experience a Mean-Time-Between-Failure (MTBF) of 3 months on a centrifugal pump. We find that various causes that include a change of service within the past year, a new bearing manufacturer is being used, and the lubrication task has been shifted to operations personnel. We take corrective actions to properly size the pump for the new service, ensure that the new bearings are appropriate for the new service, and monitor the lubrication tasks to confirm that they are being performed in a timely manner. With all these changes, we now must measure their effectiveness on the bottom line. We knew we had an undesirable situation when the MTBF was 3 months; we should now measure the MTBF over the next year. If we are successful, then we should not incur any more failures during that time period due to the causes identified in the RCA. The bottom-line effect should be that savings are realized by man-hours not expended on repairing the pump, materials not used in repairing the pump, and downtime not lost due to lack of availability of the pump.

2. *Operational*—We experience an excessive amount of rework (8%) due to production problems that result in poor quality product that cannot be sold to our customers. We find as a result of our RCA that we have instrumentation in the process that is not capable of handling a recent design modification. We also find there are inconsistencies from shift to shift in the

way the same process is operated. These inconsistencies are the result of no written operating procedures. We implement the corrective action of installing instrumentation that will provide the information we require and writing a new operating procedure that ensures continuity. Rework started at 8%, so after we implement our solutions, we should monitor this metric and make sure it comes down significantly. The bottom-line effect is that if we are reducing rework by 8%, we should be increasing salable products by an equal amount while not incurring the costs associated with rework.

3. *Customer Service*—We experience a customer complaint rise from 2% to 5% within a 3-month period. Upon conclusion of the RCA, we find that 80% of the complaints are due to late deliveries of our product to our clients' sites. Causes are determined to be a lack of communication between purchasing and the delivery firm on pickup times and destination times. We also find the delivery firm needs a minimum of 4-hour notice to guarantee on-time delivery and we have been giving them only 2-hour notice on many occasions. As a result, we have a meeting between the purchasing personnel and the dispatch personnel from the delivery firm. A mutually agreed-upon procedure is developed to weed out any miscommunications. Purchasing further agrees to honor their agreement with the delivery company in providing a minimum 4-hour notice. Exceptions will be reviewed by the delivery firm but cannot be guaranteed. The metric we could use to measure success will be the reduction in customer complaints due to late deliveries.

4. *Safety*—We experience an unusually high number of incidents of back sprain in a package delivery hub. As a result of the RCA, we find causes such as lack of training in how to properly lift using the legs, lack of warming up the muscles to be used, and heavy package trucks being assigned to those not experienced in proper lifting techniques. Corrective actions include a mandatory warmup period prior to the shift start, attendance at a mandatory training course on how to lift properly, passing of a test to demonstrate skills learned, and modifying truck assignments to ensure experienced and qualified loaders/ unloaders are assigned to more challenging loads. Metrics to measure can include the reduction in the number of monthly back sprain claims and the reduction in insurance costs and workman's compensation to address the claims.

The pattern of metric development described shows the metric that initially indicated that something was wrong can also be (and usually is) the same metric that can indicate that something is improving. Sometimes this phase seems too simple, and therefore, it cannot be used. Then we start our "paralysis by analysis" paradigms and develop complex measurement techniques, which can be overkill. Not to say they are never warranted, but we should be sure to not complicate issues unnecessarily.

EXPLOITING SUCCESSES

If no one knows of the successes generated from RCA, the initiative will have a tough time moving forward and the organization will not benefit from the effects of the analyses. Like any new initiative in an organization, skepticism abounds as to its survival chances. We discussed earlier the "program-of-the-month" mentality that is likely to set in after the introduction of such initiatives. To combat this hurdle, we need to exploit successes from RCA to improve the chances the initiative will remain viable and accepted by the work population. Without this participation and acceptance, the effort is often doomed.

How do we effectively exploit such successes? One of the main ways we do this for our clients is through high-exposure mediums. High-exposure mediums include such media as report distribution, internal newsletters, company intranets, presentation of success at trade conferences, written articles for trade publications, and finally, exposures in texts such as this one with successes demonstrated using case histories. Exploitation serves a dual purpose—it gives recognition to the corporation as a progressive entity that utilizes its workforce's brainpower, and it provides the analyst and core team recognition for a job well done. This will be the motivator for continuing to perform such work. Without recognition, we tend to move on to other things because there is no glory in this type of work.

Let's explore the different media we just mentioned.

1. *Report Distribution*—As discussed in the reporting section, to optimize the impact of RCA, the results must be communicated to the people who can best use the information. In the process of doing this, we are also communicating to these facilities that we are doing some pretty good work in the name of RCA and that our people are being recognized for it.

2. *Internal Newsletters*—Most corporations have some sort of a newsletter. These newsletters serve the same purpose as a newspaper—to communicate useful information to its readers. Most publishers of internal newsletters, with whom we have never dealt, would welcome such success stories for use in their newsletters. That is what the newsletter is for; therefore, we should take advantage of the opportunity.

3. *Presentations at Trade Conferences*—This is a great form of recognition for both the individual (and team) and the corporation. For some analysts, this is their first appearance in a public forum. While some may be hesitant at the public speaking aspect of the event, they are generally very impressed with their ability to get through it and receive the applause of an appreciative crowd. They are also more prone to want to do it again in the future. Trade conferences thrive on the input of the companies involved in the conference. They are made up of such successes, and the conference is a forum to communicate the valuable information to others who can learn from it.

4. *Articles in Trade Publications*—As we continue along these various forms of media, the exposures become more widespread. In speaking of trade publications, we are talking about exposure to thousands of individuals in the circulation of the magazine. The reprints of these articles tend to be viewed as trophies to the analysts, who are not used to such recognition. As a matter of fact, when we have such star client analysts who have written an article of their success, we frame the reprint and send it to the analyst for display in his or her office. It is something the analyst should be proud of as an accomplishment in his or her career.

5. *Case Histories in Technical Text*—As you will read in the remainder of this text, we solicited a case study from a client of ours, interested in letting the general public know of their progressive work in the area of RCA and how their workforce is making an impact on the bottom line. As most any corporation will attest, no matter what the initiative is or what the new technology may be, without a complete understanding by the workforce of how to use the new information and its benefits (personally and for the corporation), it likely will not succeed. Buy-in and acceptance produce results—not intentions or expectations of the corporation.

CREATING A CRITICAL MASS

When discussing the term "critical mass," we are referring not only to RCA efforts but the introduction of any new technology. It has been our firm's experience in training and implementing RCA efforts over the past 35 years, if we can create a critical mass of 30% of the people on board, the others will follow.

We have beat to death the "program-of-the-month" mentality, but it is a reality. Some people are leaders and others are followers. The leaders are generally the risk takers and the ones who welcome new technologies to try out. The followers are typically more conservative people who take the "let's wait and see" attitude. They believe that if this is another "program-of-the-month," they will wait it out to see if it has any staying power. These individuals are those who have been hyped up before about such new efforts, and possibly even participated in them, and then never heard any feedback about their work. They are alienated about "new" thinking and the seriousness of management to support it.

We believe if we can get 30% of the trained RCA population to use the new skill in the field and produce bottom-line results, then RCA will become more institutionalized in the organization. If only 30% of the analysts start to show financial results, the dollars saved will be phenomenal—phenomenal enough to catch executives' eyes where they continue to support the effort with actions, not words. Once the analysts start to get recognition within the organization and corporation, others will crave similar recognition and start to participate.

We have found it unrealistic to expect that everyone we train will respond in the manner that we (and the organization) would like. It is realistic to expect a certain percentage of the population to take the new skills to heart and produce results that will encourage others to come on board.

CONCLUSION

Let's face the facts—we are a human species and we are evolving. We will never be perfect, but that should not preclude us from striving to be so. We will never be error-free, but we can strive to be. Precision is a state of mind and requires the mentality to constantly strive for the next plateau.

RCA as described in this text is not a panacea. It is merely a method to assist in logical thinking to resolve undesirable events.

While many of our analogies have been from the industrial world where our background lies, we hope this RCA approach is applicable under any circumstances. Whether it is chronic or sporadic, mechanical or administrative, or in an oil refinery or a hospital, all require the same logical human thought process to resolve their respective issues.

Finally, we will show the "bottom-line results" achieved by those who had the courage to adhere to the PROACT® discipline and produce phenomenal results for themselves and their companies.

7

CASE HISTORY

This chapter puts into practice what this text has described in theory thus far. We have described in detail the Root Cause Analysis (RCA) method and provided some academic examples to further your understanding of the concepts.

The following case study is a result of having the right combination of management support, the ideal RCA team, and proper application of the RCA methodology. Return on Investment (RCI) commends the submitters of this case history for their courage in allowing others to learn from their experiences. This corporation and their RCA efforts have proven what a well-focused organization can accomplish with the creative and innovative minds of its workforce.

As you read through the summary of this actual case history, you will notice that the ROIs for eliminating these chronic events are expressed in the thousands of percent. Had we not had permission to publish these remarkable returns, would anyone have believed they were real? While these results are without a doubt impressive, they are easily attainable when the organizational environment supports the RCA activities. Read on and become a believer.

CASE STUDY: NORTH AMERICAN PAPER MILL

Undesirable Event: Repetitive Thick Stock Pump Failures

Undesirable Event Summary: During the years 2007 and 2008, there were 19 failures on thick stock pumps on A and B units in the Bleach Plant. Several attempts had been made to implement corrective actions for the pumps, but ultimately, failures were still occurring. Thick stock pumps are big-ticket items ranging from $60,000 to $120,000 per rebuild due to the tight clearances and amount of material it takes to machine the pumps. It was determined by maintenance that the pumps could be rebuilt in-house in the bleach room maintenance shop. This has been very successful and has cut the cost of maintenance dramatically and has proven to provide greater reliability. Performing the rebuild by in-house millwrights has brought ownership and pride to the repairs and operations of the thick stock pumps. Although production loss was not used in the original "Opportunity Analysis" for these failures, it would have been a significant factor in the loss equation for these events.

Through the RCA, the team determined the two prevalent causes for premature failure were bearing failures and foreign material being introduced into the pumps. Foreign material is an issue that persists and is a random failure. This occurs because the clearance (0.03) in the pumps is very small and does not allow any metals bigger than a paper clip to be passed through. With closer attention to this issue, there have not been any failures due to foreign objects in the stock reported since this RCA was completed.

The RCA team focused on bearing failures during the RCA, which led to many preventive actions that were recommended and implemented. Several issues had been identified that caused contamination to find its way into the lubrication chambers. Packing was the first area that was pursued. New flow visual indicators and check valves on the packing water lines helped to ensure enough packing water was getting to the packing. A new procedure was recommended to properly pack a thick stock pump and was later implemented. To monitor how often the lubrication becomes contaminated, a predictive maintenance technique (lube oil analysis) is now being utilized. Oil analyses are being performed on all thick stock pumps every 2 weeks in order to determine how often the oil requires changing. This will be sustained by putting it on the area oiler's checklist.

Also, the team implemented both operator and maintenance ECCM (essential care/condition monitoring) routes to monitor the operation of these critical pumps. Operators perform visual inspections of the thick stock pumps and motors, in addition to recording vibration, temperature, and pressure readings into handheld data collectors for early detection of defects. This is in addition to other more advanced vibration analyses that take place.

Identified Physical Roots
Lubrication issues due to contamination
Pack failures
Packing water failures (plugged, inadequate pressure, etc.)

Identified Human Roots
Inadequate packing
Inspections not performed
Poor rebuilds from outside services

Identified Latent Roots
Using the wrong packing procedures
No formal inspection routine for operations and maintenance
Inability to determine if packing lines were in working order
Assumed that repairs were performed properly by OEM

Implemented Corrective Actions
Improved packing procedure implemented and documented in
   SAP Plant Maintenance (SAP PM).
Oil analysis implemented.
Development and execution of ECCM inspection routes.
New visual flow indicators installed for packing water lines.
Improved check valves installed for all thick stock pumps.
Defined all new parts as storeroom items for future use.
Perform in-house rebuild on thick stock pumps.

Effect on Bottom Line
Tracking Metrics
Mean-Time-Between-Failure (MTBF) increased from approxi-
   mately 6 months to over 2 years.
Maintenance cost reduced less than 25% of pre-RCA costs.

Bottom-Line Results
There are substantially fewer failures on these critical pumps, and
   defects are caught before they cause catastrophic problems.
In-house rebuild of pumps has resulted in greater ownership and
   pride in the performance of the pumps.
Maintenance costs in 2010 YTD (July) are approximately $25,000
   compared to roughly $500,000 in 2006 and 2007 in the same
   time frame.

RCA Team Statistics
Start Date: July 30, 2007
End Date: October 20, 2008
Estimated Cost of Performing the Analysis: $20,000
Approximate Savings: $700,000 per year
Estimated ROI: 3,500%

Corrective Action Time Frame

Most of the corrective actions were implemented in less than 6 months.

The analysis spanned about 3 months.

The implementation of corrective actions was complete approximately after the completion of the analysis.

Core Team Members

Arnie Persinger (Principal Analyst)
Dean Muterspaugh
Todd Fix
Robert Newcomer
Josh Taube
Rob George
David Persinger
Will Sales

Special thanks to Arnie Persinger (Area Maintenance Superintendent) and Craig Lane (Fiberline Operations Superintendent) for believing in the process and ensuring a successful outcome.

INDEX

Printed in the United States
by Baker & Taylor Publisher Services

Printed in the United States
by Baker & Taylor Publisher Services